THE PANTANAL OF MATO GROSSO (BRAZIL)

MONOGRAPHIAE BIOLOGICAE

VOLUME 73

Series Editors
H.J. Dumont and M.J.A. Werger

The titles published in this series are listed at the end of this volume.

The Pantanal of Mato Grosso (Brazil)

World's Largest Wetlands

by

F. D. POR

Department of Evolution, Systematics and Ecology
The Hebrew University of Jerusalem
Israel

Springer-Science+Business Media, B.V.

A C.I.P. Catalogue record for this book is available from the Library of Congress

ISBN 978-94-010-4018-1   ISBN 978-94-011-0031-1 (eBook)
DOI 10.1007/978-94-011-0031-1

Printed on acid-free paper

All Rights Reserved
© 1995 Springer Science+Business Media Dordrecht
Originally published by Kluwer Academic Publishers in 1995
Softcover reprint of the hardcover 1st edition 1995
No part of the material protected by this copyright notice may be reproduced or
utilized in any form or by any means, electronic or mechanical,
including photocopying, recording or by any information storage and
retrieval system, without written permission from the copyright owner.

Table of Contents

Preface vii

1. Introduction 1
2. Discovery 3
3. Limits and size 5
4. Geological data 8
5. Local geomorphological terminology 14
6. The climate 16
7. River hydrology 19
8. The Amazonian connection 22
9. The flood regime 23
10. The Pantanal lakes 28
11. Hydrochemistry 31
12. Soils 34
13. How many Pantanals? 34
14. Regional connections 37
15. Introduction to the biology of the Pantanal 38
16. Phytogeography 41
17. Aquatic vegetation 44
18. Terrestrial vegetation 50
19. Fire in the Pantanal 57
20. Zoogeography 58
21. Aquatic meiofauna 60
22. Macrobenthos 64
23. Ichthyofauna 65
24. Pantanal fisheries 71
25. The herpetofauna 72
26. The avifauna 77
27. The mammals 84
28. The humans in the Pantanal 89
29. Environmental worries 98
30. Nature conservancy and ecotourism 100

Taxonomic glossary 111
Geographical and ethnographic index 118
Subject index 120

Preface

For a naturalist and limnologist, the Pantanal has the extreme fascination of an "ultima Thule" of undisturbed and little known wilderness. The scientific world at large is almost unaware of its richness. In an age when scientific research is overstretched because of lack of funds and the hands are full of urgent conservation tasks, it is the amateur tourist who unveils the beauty and the interest of this largest wetland of the world.

I had the privilege of an outsider, well enough familiarized with Brazil, its language and scientific life, but independent enough of the daily chores of a local academic career. For nearly 20 years I have been a faithful scientific tourist to this subcontinent. My academic liberty gave me the unique opportunity to try to synthesize in English the knowledge about Brazilian environments little known abroad. My first such endeavour has been "Sooretama – the Atlantic Rain Forest of Brazil". Dealing now with the Pantanal, I wish to pay tribute to the many Brazilian colleagues who under dire and often precarious conditions have advanced the knowledge related to the Pantanal. By reviewing their many reports and papers written in Portuguese and bringing them to the knowledge of the international scientific community, I believe that I am doing a useful service.

First of all, I have to thank Prof. Vera Lucia Imperatriz Fonseca, the Head of the Department of General Ecology at the University of São Paulo, for providing me during the years, the academic base from which I could operate. Thanks are due to Dra. Nicia Wendel Magalhães who awoke my interest for the Pantanal and organized my first trip to that area. All my appreciation goes also to colleague and friend Prof. Carlos Eduardo Falavigna Rocha, Department of Zoology, University of São Paulo, with whom I collaborate for several years on many copepodological subjects and now enthusiastically joined our Pantanal project. Finally, Prof. Antonio Carlos Marini from the Federal University of Mato Grosso do Sul in Campo Grande, André de Castro C. Moreira from the Life Sciences Institute of the University of São Paulo, and Arnildo Pott and Vali Joana Pott from EMBRAPA Corumbá provided me with important information.

Adão Valmir dos Santos, then student at the UFMS, guided our first trip in which also colleagues Dra. Ana Millstein and Marcello Juanico from Haifa, Israel participated.

Mr. Peter Grossman, Jerusalem, drew and redrew all the maps and graphs which appear in this book.

Finally, I want to thank my dear wife Dra. Maria Scintila de Almeida Prado Por, who from the first hardship when our car broke down on the dirt tracks of the Pantanal and till the publication of this book, has been a permanent stimulus and support to me.

May this book serve as a small contribution to the study and preservation of this most beautiful of Brazil's environments, the Pantanal of Mato Grosso.

February 1995 F. D. POR
Jerusalem, São Paulo

1. Introduction

Near the geographical centre of South America is the largest complex of wetlands of the world (Alho et al., 1988). This is the Pantanal of Mato Grosso in Brazil, adjoining the upper reaches of Rio Paraguay. Extending also somewhat into neighbouring Bolivia and Paraguay, it is also known by its Spanish name "Gran Pantanal". The meaning of the word Pantanal, is simply "Wetland". For Eiten (1983), all the Brazilian freshwater wetlands are "pantanais". Here, however, we shall use the term in its strict geographic meaning as the "Great Pantanal" of Mato Grosso.

Even at a first sight on the map of the continent, the Pantanal appears as a natural switchboard between the two great river basins of South America, namely the Amazon and the La Plata basins (Fig. 1).

With a size of around 200,000 square kilometres (see below), this is an area comparable to an average sized European country. Unlike other large-sized wetlands of the globe, namely the some six times smaller Sudd swamps in Africa and the tens of times smaller Everglades, the Pantanal has a complicated topography. It consists of tens of large-sized rivers and their deltas, thousands of lakes and salt pans, all this interspersed with scrubby savanna and grassland, stretches of riverine forest, fragments of, mountain ranges and scores of isolated rocky monadnocks. As pointed out by several Brazilian authors, the name "Pantanal" may be misleading, because the area is not a homogeneous, monotonous swamp, although the swamps often stretch out of sight. Situated in a region of extreme continental climatic fluctuations, large stretches might be for some time completely dry and semi-arid and yet be flooded suddenly according to complex and not yet elucidated hydrographic circumstances. In the terminology used by Walter and Breckle (1984), the Pantanal is a "Parklandschaft", a park environment, a biome characterized by extreme fluctuations of humidity, tending towards a wetland. The Pantanal is an immense syncline, in which rivers, lakes and wetlands are interfingering with pioneer terrestrial ecosystems in a permanently shifting and imprevisible dynamism. Time and again, following secular or decade-scale cycles, the whole extension of the Pantanal turns into an anastomosed watery continuum.

The Pantanal is perhaps one of the least known regions of the globe. Scientific information is surprisingly scarce. Although private ranches divide among themselves the whole area, these sites can be approached as a rule only by small planes and by outboard craft, or, floods permitting on horse back or with four-wheel vehicles. There are large areas which were never properly surveyed and charted in the field and much of the information results from the interpretation of satellite photography (the RADAMBrasil programme). There are huge lakes which have never been properly fathomed, nor touched by the net of a limnologist.

In recent years, blooming tourist interest for the Pantanal, both in Brazil and from abroad developed, making some inroads into the huge area. But an understanding of the whole phenomenon of the Pantanal is dangerously

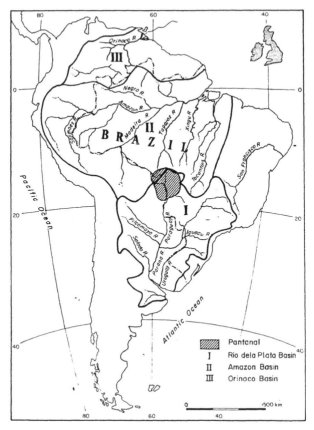

Fig. 1. Localization map of the Pantanal in South America. The three main river catchments of the Neotropics are emphasized. In the upper right corner the British Isles are drawn at scale, for comparison of sizes.

lacking. We do not really know what are the factors which made the Pantanal to be "the site with the richest faunal concentration of the world" (Sucksdorff, 1989).

There exists a score of more comprehensive works about the Pantanal, written in Portuguese. Best among them is the recent guidebook by Nicia Magalhães (1992), which I used extensively as a source of information. According to the review made by Cadavid-Garcia (1992), there are 820 publications about the Pantanal, of which 85% are technical reports in Portuguese. Some 316 publications are in the field of botany and agriculture and only 8 publications deal with invertebrates!

In the international literature, the Pantanal is almost unknown. Besides two German publications (Wilhelmy, 1958; Klammer, 1982), I could find only three English titles of wider scope: Prance and Schaller, 1982; Alho et al., 1988; Junk, 1993. The two symposia organized by the scientists of Mato

Fig. 2. The Pantanal ("Eupana Lacus") as it appears on the 1559 map by Hondius (courtesy M.C. Mesquita Correa Pereira).

Grosso do Sul, respectively, in Corumbá (1984) and in Campo Grande (1989), published as in-house publications in Portuguese, served only to expose the many lacunas in the knowledge of the Pantanal. There was a feeling that the immensity of the Pantanal and its imprevisible flood regime provide a certain guarantee against dangerous anthropic changes and that the time is still available for leisurely research. But the circumstances have changed now dramatically: a consortium of five states, namely Brazil, Paraguay, Bolivia, Argentina and Uruguay have decided to implement a 1990 agreement to regulate Rio Paraguay and to open it up to sea-going navigation, including the over 1000 km stretch in which this river crosses the Pantanal. Intensive international research, to which events such as the forthcoming International Congress of Limnology (SIL) should serve as a catalyst, will be needed in order to minimize the impact of this large-scale engineering enterprise.

2. Discovery

The legendary existence of a great inner lake in the centre of South America was believed for several centuries. On the 1559 map by Hondius, the Pantanal is represented as a large lake, "Eupana Lacus", surrounding an archipelago (Fig. 2). In the 1641 improved edition of the Hondius map, the swampy lake is

Fig. 3. A lithograph portrait of Georg Heinrich von Langsdorff at young age, by an unmentioned artist (from Braga, 1988).

still there, but no name is given. This lake appeared as the common source of the Paraguay, the São Francisco and even of the Amazon (Hoehne, 1936).

The name of the "Sea of the Xaraes" was first mentioned by the Spanish conquistador Nuñes Cabeza de Vaca, who after founding the city of Asunción, travelled in 1543 upriver till lake Gaiba. There he must have heard from the "sea" from the indian tribe of the Xaraes, which inhabited the shores of another extensive lake, Lagoa Uberaba, further north. Cabeza de Vaca who invented also the tale of the seven golden cities of Cibola in Texas, probably did not check his source too much. To his defence one can admit that a lake, the size of Uberaba, which can reach a surface of over 400 square kilometres, can easily be considered as an "inland seas" even under the present climatic conditions.

For two centuries the swamps of the Pantanal were still the domain of the indian tribes, the Payaguas, expert boatmen, the Guaicurus who became feared horsemen, and others. During the XVIIIth century, the "bandeirantes" slave-capturing armed bands from the province of São Paulo discovered gold near Cuiabá. In rapid sequence, the present frontier line separating Portuguese from Spanish lands became established and studded with stronghold townships. But the indian tribes remained a fiercely independent menace for the colonists, even during most of the XIXth century. In a typical North-American way, these powerful tribes choose either the Portuguese or the Spanish side, even as late as in the Paraguay War of 1864–1870 (Valverde, 1972; Gomes de Souza, 1986).

Scientific expeditions were late to arrive in this dangerous and inhospitable lands. An account of the botanists which carried out expeditions to the Pantanal is found in Urban (1902). The first collector in the area was Alexandre Rodrigues Ferreira who herborized in the surroundings of Cuiabá in 1790. However, the most impressive enterprise was that of the count George Heinrich von Langsdorff, between 1824 and 1829 (Fig. 3).

Langsdorff who first took part in the Russian circumnavigation of the world by Krusenstern, came to act as Russian consul in Rio de Janeiro. From there he organized an expedition composed of a botanist, two zoologists, an astronomer and two draughtsmen. It sounds now very bizarre, but the first Pantanal expedition was paid by the czar and travelled under the flag of the Russian Empire. Langsdorff's eccentric personality has been the subject of many studies, and especially since his expedition ended in near debacle, as one of the painters drowned and Langsdorff himself developed a terminal mental illness on the outgoing leg from Cuiabá (Komissarov, 1988; Pinto Braga, 1988).

Almost simultaneously with the ill-fated Langsdorff expedition, the zoological expedition of Johann Natterer amassed an important vertebrate and mainly ornithological collection from the Pantanal, this time under the orders of another emperor, that of Austria. Natterer should be considered as the real pioneer of the Pantanal research, because unfortunately most of the Langsdorff material has been probably lost or left undescribed.

During the remaining years of the XIXth century only the expedition of Alcide d'Orbigny and of Charles Godichaud are on record. The geographer Carl von Steinen passed through the region, but supplied little more than geographic data about the northern edges of the Pantanal.

Till late in the last century, the name used for the region has been the old Portuguese "Melgaço" meaning badland or wetland. The name "Pantanal" became common in literature through the report of the expedition of Marchant Moore (1890–1893) (Padua Bertelli, 1988).

Worth mentioning is the expedition headed by the former president Theodore Roosevelt, which passed through the Pantanal in 1913. Some of the pioneering scientists of modern Brazil took part in the Roosevelt expedition, such as the botanist Hoehne, the herpetologist medic Vital Brasil, the parasitologist Oswaldo Cruz and the ethnographer-politician and later marshal Cândido Rondon, a native of the Pantanal.

The first modern information resulted late in the present century with the advent of aerial photographs and with the establishment of a research institute in the towns bordering the Pantanal. Without any significant international research, the Pantanal is still a scientific virgin land, in which the Brazilian scientists only ploughed the first furrows.

3. Limits and size (Fig. 4)

Situated roughly between 16° and 20° S and 58° and 50° W, the limits of the Pantanal are not unanimously accepted. In essence it is an intracontinental basin (Almeida, 1945), with an elevation between 200 and 75 m, of a pan-shaped core area delimited mostly by river Paraguay in the west and the

Brazilian uplands in the east and a gradually narrowing panhandle to the south. The core area with some 130,000 square kilometres has extensions to the south, the north and to the west across the Paraguay. The limits of the Pantanal towards the Bolivian Chaco are not well defined and known. In the north, the Pantanal connects with the marshy lowlands of Rio Guaporé. Because of these incertitudes, the extension of the Pantanal appears widely variable in the different publications: from 250,000 square kilometres (Tundisi and Matsumura-Tundisi, 1985) to mere 80,000 square kilometres (Bonetto, 1975). Considering that about 10% of the Pantanal extend into Bolivia and Paraguay, the total extension of the Pantanal is probably 200,000 square kilometres, which corresponds with the figure given by Couto et al. (1975). Waiter and Breckle (1984), include all the catchment of the Paraguay in the area of the Pantanal and reach a figure of nearly 500,000 square kilometres, a view which in a historical sense might be correct. In recent conservationist Brazilian literature (see for instance Coutinho et al. (1994), the whole Brazilian part of the catchment of the Upper Paraguay is considered, i.e., 393,600 square kilometres, of which 140,000 are floodplains and 253,600 kilometres are high plains. This is without doubt a correct approach if one plans the preservation of the whole basin.

If the Pantanal is an area characterized by a certain climatic regime (see below), than the idea of Dourojeani (1988), to see the Pantanal as being circumscribed by the isohyet of 1300 mm is a very interesting approach.

The Pantanal can be considered to be the base-level collector of the basin of the Upper Paraguay. The catchment of the Upper Paraguay is of nearly 490,000 square kilometres, of which some 380,000 are situated in Brazilian territory (Oliveira Carvalho, 1986). However, almost all the tributaries from the Bolivian and Paraguayan Chaco are temporary torrents.

The Pantanal basin has an average width of 300 km with a maximum of 500 km. The maximal longitudinal transect is also of about 500 km. The eastern border is clearly set by the Brazilian highlands situated at about 400 m altitude. This eastern border hills are called Serra do São Jeronimo in the north and Serra do Maracaju in the south. The escarpment closes the Pantanal also from the south under the name of Serra da Bodoquena. As the lowlands are gradually compressed between the Serra da Bodoquena and the hills on the Bolivian side in the west, the Pantanal forms a narrow panhandle limited to the south by Rio Apa, a tributary of the Paraguay. At about the same level the Paraguay itself crosses the narrows between the mountains of the two banks in an area called "Fecho dos Morros" (The Closure of the Hills). This is considered also the limit of the Upper Paraguay basin.

The escarpments of the Brazilian highlands define also the northern border of the Pantanal, under the names of Serra do Chocororé, Chapada dos Guimarães and Chapada dos Parecis. There is an opening to the north in the Gate of Coxim, through which the Paraguay forces its way southward. In this

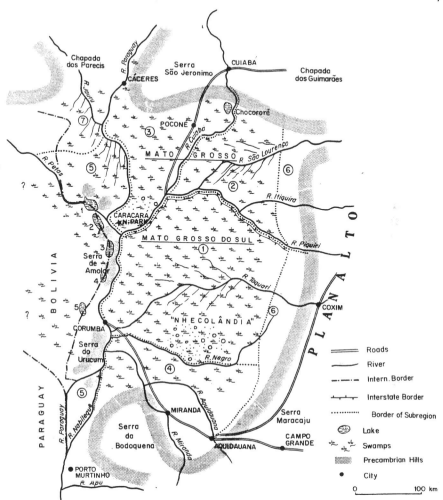

Fig 4. An orientation map of the Pantanal. The subregions follow Klammer (1982): 1. Alluvional fan of Rio Taquari; 2. Fan of Rio São Lourenco; 3. Sand alluvions of Rio Cuiabá; 4. Sediments of Rio Miranda and Rio Aquidauana; 5. Varzea of Rio Paraguay; 6. Pediments of Taquari and São Lourenco; 7. Alluvional fans of Rio Jauru and Rio Paraguay. Lakes are not drawn to scale. Location of the Pantanal National Park is indicated by asterisk.

area the marshlands of the Pantanal extend way north, behind the barrier of the hills.

The limit to the west, beyond the Paraguay and among the high ground of the Bolivian Chaco, is basically a sharp climatic gradient of increasing aridity. The basins and lakes turn into endorheic saltpans and the rivers into temporary wadis.

On its way south, the Paraguay is contained by a row of short and steep Precambrian ridges situated on its western shore (Fig. 5) In a few instances

Fig. 5. Rio Paraguay and its swamps following the range of the West bank Precambrian hills (Dubs, 1992).

these ridges reach altitudes of over 1000 m. Alternating with the mountain chains, there is a sequence of large lakes, which are partly collectors of floods coming from the Chaco and partly overflow lakes of the Paraguay and of its large left bank tributaries. Along some of the intermittent right bank rivers, like Rio das Petas (Coricha Grande) which feeds lake Orion and Uberaba, and Rio Otoquis in the southern corner, the swamps penetrate finger-like for some distance into the Bolivian and Paraguayan lands. The swamps of Otoquis collect the waters of the equally named river which has a large catchment area of 18,000 square kilometres but is, as mentioned, a seasonally flowing river.

4. Geological data

The short and rather sketchy data below, follow Almeida and Lima (1956), Braun (1977), Klammer (1982), Tricart (1986), Godoi Filho (1986), Ab'Saber (1988) and Petri and Fulfaro (1988).

The basin of the Pantanal is an old continental margin of the pre-Cretaceous South American Gondwana. In the literature there are many speculations about the existence in the area of a marginal sea or of a large lagoon of the Pacific. Grabert (1964) and Putzer (1984) speak of a Neogene brackish lagoon, gradually isolated from the ocean by the uplifting of the Andes (Fig. 6). According to Petri and Fulfaro (1988), the Paraná basin underwent

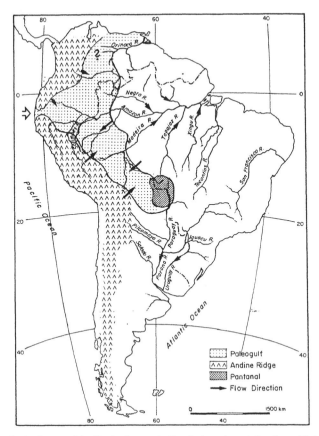

Fig. 6. Hypothetical map of the Early Cenozoic South-American inland sea. Note the drainage to the Pacific through the "Guayaquil Gap".

a tectonic subsidence around 4.6 million years ago, i.e., at the end of the Eocene, possibly related to the uplifting of the nearby Andes. To what extent the Pliocene brackish formations known from the Upper Amazonas and the Acre basin under the name "Pebas formation" were contiguous with the Pantanal basin is not reported (Petri and Fulfaro, 1988).

There is no fossil documentation for the older brackish or lacustrine history of the Pantanal. In fact only old Precambrian-Cambrian fossils have been found in the area. Such is the material of *Cloudina*, worm-like animals, found by Fairchild (1978) in the Serra do Urucum. If there has been indeed a later marine or brackish gulf in the site of the Pantanal, the proofs for this are buried under hundreds of metres of thick Pleistocene sediments.

The surrounding escarpments, mountain ridges and monadnocks are all of precambrian or early Palaeozoic age. They are locally rich in iron and manganese ore and contain also widespread calcareous rock. Karstic phe-

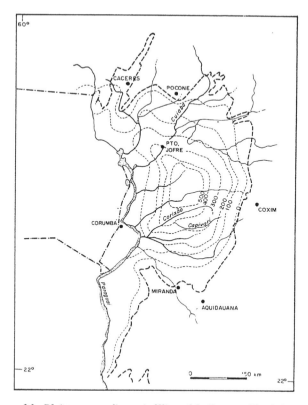

Fig. 7. Thickness of the Pleistocene sediment infilling of the Pantanal basin in meters (redrawn after Godoi Filho, 1986).

nomena are frequent with a wealth of caves and dolina's. Already Hoehne (1936) discussed extensively the calcareous rocks of the Pantanal borders. Tricart (1982) mentions the abundant karstic spring of Fonte Progresso in the north, while in the south the headwaters of Rio Miranda are formed by several hypercalcic tributaries, such as for example Rio Formoso, near Bonito.

The ancient hills may be remnants or peaks of a "circum-Pacific land or island belt, which may once have lain in front of the whole Pacific coast of America" (Fittkau, 1969). This would be the hypothetical "Archiplata" of the German authors (see below).

There is today a growing acceptance that the graben of the Pantanal itself was formed through the fragmentation of an old Cretaceous shield, in a balancing conjunction with the uplifting of the Andes and of the Brazilian highlands. the topography changed from a basically exorheic pre-Tertiary domed shield, to an endorheic syncline. As the orogenesis progressed, the syncline of the Pantanal sunk deeper and deeper and filled up with alluvional sediments. Though not well substantiated, there is every reason to believe,

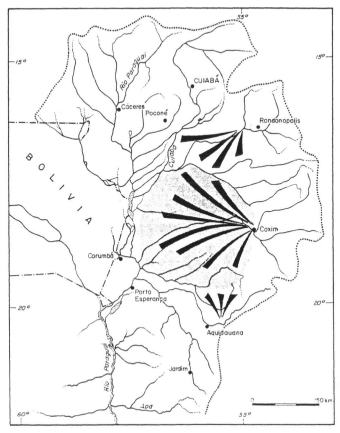

Fig. 8. Brazilian drainage of Rio Paraguay, with the alluvional cone of Rio Taguari and other cones (redrawn after Braun, 1977).

that the Pantanal depression was indeed filled time and again by a paleo-lake or by several lakes.

According to Godoi Filho (1986), the underlying basement of the alluvional fill is constituted by the Plio-Pleistocene travertines of the Xaraes formation. Already in 1894, Evans characterized the Xaraes chalks as lacustrine sediments and reported fossil freshwater gastropods of the genus *Ampullaria* in them. The gradual subsidence of the Pantanal graben has been a protracted process. The drillings carried out by Brasil's national petrol company Petrobras, found maximal sediment infillings of about 500 m thickness in the very middle of the Pantanal (Fig. 7). This, means that the bottom of the graben is now sunken more than 400 m below sea level (Ab'Saber, 1988).

The Pantanal, as we see it today, represents the fluctuating history of the Pleistocenic climates. During the arid glacials, the area turned into a waste endorheic basin of the Chaco type, a pediment over which seasonal torrential rivers deposited large alluvional fans. Rio Paraguay itself ran out into such an alluvional cone, more or less at the level of the town of Cáceres. During the alternating humid Interglacials, the rivers acquired a permanent flow regime and flew over the alluvional cones. River Paraguay turned into a collector which succeeded, probably time and again, to leave the depression over the sill of the Fecho dos Morros (Tricart, 1986). This is more or less the present situation.

It results that the Pantanal consists basically of the alluvional fans of its main rivers (Braun, 1977; Klammer, 1982) the standing water bodies on these fans and the deltaic network of streams which criss-cross the old fans. By far the most important fan is that of Rio Taquari, probably the largest of its kind in the world. It has a surface of 50,000 square kilometres (Fig. 8). Smaller alluvional fans are those of Rio Paraguay, jointly with its tributary Rio Jauru in the north, the combined fans of Rio Cuiabá and Rio São Lourenço and in the south the joint fans of the rivers Miranda and Aquidauana. Studying the paleo dunes which accompany the old alluvional fans, Klammer (1982) reaches the interesting conclusion that the Pantanal passed only through one protracted arid phase, sometime in the latest pre-glacial Pliocene. This contrasts Ab'Saber (1988), which in the spirit of the Pleistocene refugia hypothesis, believes of a fluctuating repetition of the humid and arid phases during the Pleistocene.

The alluvional fans pushed the present bed of the Paraguay towards the line of the Precambrian escarpments. In the north of the graben, where this restriction does not exist, Rio Paraguay flows through a very broad alluvional valley.

During the arid phase, the Pantanal was probably a "paleo desert" (Klammer, 1982), strewn with dunes and occupied by intermittent, locally resurgent temporary streams and many salty terminal lakes. Ab'Saber (1988) considers that the last period of drought occurred only 23,000–13,000 years ago, during the last glacial paroxysm.

According to the CLIMAP maps (Brown, 1986) the temperatures of the last pleniglacial were on the average 8 °C lower than today. Probably the effect of the Antarctic fronts, felt even today, was much more massive.

The present features of the lakes and marshlands have been established probably after the maximal humid phase of the "Postglacial Climatic Optimum", 7000–5000 years ago. During the humid optimum the overflow-lakes in the Pantanal were probably considerably larger and may have survived in the memory of the aborigine populations as the fabled "Sea of the Xaraes". Veloso (1972) considers, based on phytogeographic evidence, that the Pantanal passes presently through a phase of gradual drying, a process which is accelerated by human interference.

Under the present climatic conditions, the gallery forest vegetation overgrew the banks of the rivers and retains the sediments carried by them.

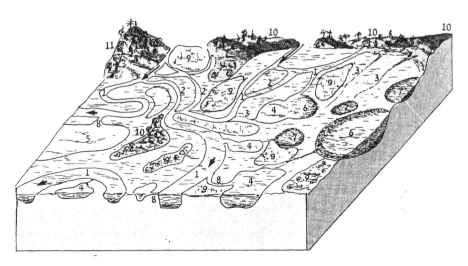

Fig. 9. A scheme of the toponymy of the main physical features of the Pantanal. 1. "rio"; 2. "corixo"; 3. "vazante"; 4. "baia"; 5. "lagoa"; 6. "salina"; 7. "barrreiro"; 8. "boca" or "sangradouro"; 9. "banhado" or "vàrzea"; 10, "cordilheira"; 11. "serra" or "morro".

Fig. 10. The lower course of Rio Miranda.

Fig. 11. Landscape of Nhecolândia, with round baías and salinas (Anonymous, album "Pantanal é Vida").

Therefore further formation of alluvional features is prevented. (Klammer, 1982; Tricart, 1982).

5. Local geomorphological terminology (Fig. 9)

A complex environment, like that of the Pantanal, is bound to generate a variety of local terms, by which the "pantaneiros" define their surroundings. Since these terms are not strictly codified, they may be interchangeable and locally different.

Rivers with defined headwaters ("cabeçeiras") are called "rios" (Fig. 10). Braids which connect rivers, are in general called "corixos" and smaller ones "corregos". The same terms are in use also for the seasonal torrents of the Bolivian side. Temporary flood drainages are called "vazantes". Deltaic outflows into a main collector are "vertentes". Short an relatively deep channels which connect lakes with rivers are "sangradouros" or "bocas". This names may differ from place to place, probably expressing also the permanent change in the hydrological features.

Fig. 12. View of the meanders of Rio Miranda, from morro do Azeite.

Fig. 13. A small and overgrown baía, with a cordilheira in the background and isqueiros at work.

Lakes are called "lagoas" when large and "baías" when smaller. "Salinas" are the salt lakes and "lambedouros" (licking places) and "barreiros" are the drying out sebkhas (Fig. 11). With the building of roads through the marshes, a new type of roadside lakes appeared, the "caixas de empréstimo".

Fig. 14. Morro do Azeite on Rio Miranda.

"Varzeas" and "baixadas" are seasonally flooded areas, locally called also "largos", "várzeas" and "banhados". Dry lands are also of several types: The "diques" are the risen banks of the rivers, bearing gallery forests (Fig. 12). These are called locally "saranzais". The "cordilheiras" or "capões" are stripes of dryland usually old dunes which often delimit baías (Fig. 13). The cordilheiras reach on average a height of 4 m and are therefore protected from normal floods. "Morros" and "morrarias" are isolated Precambrian hillocks, usually a few hundred metres high (Fig. 14). Often they are surrounded also by gently rising slopes, the "fraldas". Whole ridges of such old mountain formations are the "serras" or "serranias".

In the following, I shall frequently use the local terminology.

6. The climate

The climate of the Pantanal is of the "Aw" type in Köppen's classification, i.e., with the dry winters (May to September) and wet summers which characterize the savanna's of the southern hemisphere. The annual temperature media fluctuate around 25 °C (Fig. 15). The highest temperatures occur usually

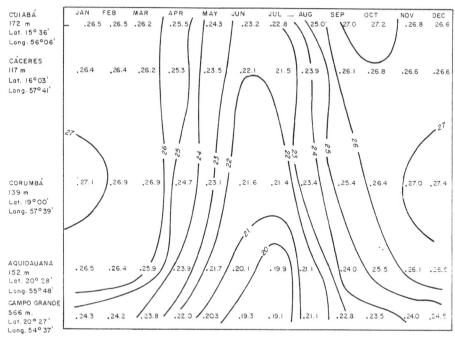

Fig. 15. Spatial distribution of the mean annual isotherms between the years 1931–1960) (from Tarifa, 1986). Note that Campo Grande is on the high ground of the Planalto.

early in the summer and may reach 40 °C. According to Valverde (1972), the wet summer regime is due to the penetration of the equatorial continental air mass of Amazonian origin. The climate of the winter is dominated by the tropical-atlantic air mass coming from the Brazilian highland. Since the amphitheatre of the Pantanal is open to the south, sometimes polar-antarctic atmospheric fronts advance into the area and winter temperature extremes of around 0 °C may occur. These are the so-called "friagens" which can provoke frost-bite etiolation of the plants over large swamp areas.

Finally, the hot and dry air mass which forms over the Chaco in the summer, does not penetrate presently the main area of the Pantanal, but may influence the southern panhandle around "Rio" Nabileque and Rio Apa, the driest part of the Pantanal.

Humidity is usually around the mark of 70%, reaching maxima of over 80% in the late summer (Tarifa, 1986). According to the most recent study by Alfonsi and Paes de Camargo (1986), most of the Pantanal is characterized by an annual hydrological deficit of 300 mm, resulting from the medium pluviosity of somewhat below 1,100 mm and an annual evapotranspiration of around 1,400 mm (Figs. 16 and 17). The surrounding highlands, the annual precipitation is well over 1,500 mm and there is no evaporation deficit. This increases dramatically over the Bolivian Chaco.

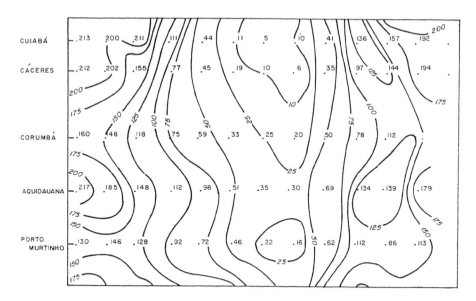

Fig. 16. Spatial distribution of the monthly rainfall (in mm) (after Tarifa, 1986).

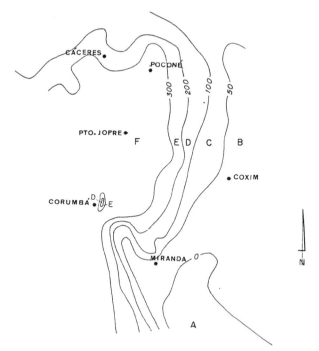

Fig. 17. Distribution of the annual hydrological debt (net water loss in mm), over the Pantanal (modified after Alfonsi and Paes de Camargo, 1986).

The Pantanal therefore, is a large climatic enclave in which the run-off from the surrounding relatively wet highlands, carried by a series of large rivers, succeeds to maintain an allochthonous wetland environment under the conditions of a basically semi-arid climate. Klammer (1982) wrote that the common feature of the Pantanal rivers is the fact that their flow decreases downstream, instead of increasing as usual.

The Paraguay itself, is to this author an allochthonous river.

From a different point of view, the Pantanal is probably the most important window of evaporative freshwater loss of the globe. This is probably somewhere in the range of 60,000 cubic kilometres/year if one considers a total surface of 200,000 square kilometres and a net evaporative loss of 300 mm/year. Calculating the difference between the inflow and the outflow of the Rio Paraguay in a defined sector, Pfaffstetter (1974) accounted for the loss of 250 million cubic metres in the hydrological year 1968–1969. All these calculations deserve a more precise mathematical model which would take into account also the losses of water to the groundwater, as well as the long-term retention of water by the plant biomass.

Since the wetlands of the Pantanal owe their existence to the waters of allochthonous rivers, rather than to local pluviosity, the whole huge area is subject to local and episodic floods which depend on hydrological conditions at the river headwaters, far and outside the Pantanal itself. Typically for an endorheic basin, the Pantanal presents the driest climate in the lowest lying areas, in the west and southwest. These are also the most frequently flooded ones.

7. River hydrology

The Pantanal depression is extremely flat-bottomed. From north to south, along the axis of Rio Paraguay, the declivity is of only 37 cm/km and this low value decreases towards south: In fact the Pantanal basin is slightly rising as one approaches the shallow outflow of the Paraguay at the Porto Murtinho outflow. Along the east-west axis the declivity is in the order of 13.6 to 35 cm/km. This explains in great part the fact that the fluviatile network of the left bank tributaries is extremely complex and permanently shifting. In fact we are dealing with a complex of several internal deltas, the branches of which are activated and deactivated by the smallest changes in flow rates, sedimentation rate and topographic adjustments. The main river beds, well defined in their upper reaches become confuse as they approach the Paraguay. As a consequence, even the nomenclature of the lower branches is not unanimously defined. The local fishermen often disagree in deciding which is the real river and which a corixo or a vazante.

Rio Paraguay takes its origin on the highlands of the Chapada dos Parecis and its upper reaches the right side tributaries Sepetuba, Cabeçal and Jauru.

The Jauru itself is a major river with diffuse headwater connections with the Amazon tributaries (see below). The uppermost stretch of the Paraguay has a flow of about 440 m^3/s. Upon emerging below Cáceres into the Pantanal lowlands, the floodplain widens from around 5 km to many tens of kilometres. The main talweg of the river becomes less defined as it flows upon its old alluvional cone and passes through the large lakes of Uberaba and Gaiba. further south along its middle stretch, the river is pushed against the western mountain ridges by the deltaic accumulations of the left side tributaries. After the inflow of Rio Miranda, the Paraguay widens again and becomes more deltaic.

The river presents several deltaic branches or old river beds. The most important ones are Paraguai Mirim ("Little Paraguay"), 120 km long and Rio Nabileque, 250 km long. At high waters, the whole width of 90 km between the Paraguay and the Nabileque is under water. Ab'Saber (1988) compares the Nabileque to a "misfit river", i.e., a river which is much smaller than the valley through which it presently flows. Thus, it is possibly a deserted river bed of the Paraguay.

Upon leaving the Pantanal through the Fecho dos Morros gate, the Paraguay has a flow of 2000 m^3/s. As emphasized by Bonetto (1975), Rio Paraguay in its lower stretch contributes about 25% to the total of the La Plata system, but this contribution is seasonally constant. The stretch of the river, in the Pantanal depression acts as a regulator of the fluctuating discharges of its tributaries. Compared with its sister river, the Paraná, which has no headwater swamps, the Paraguay carries also much clearer water (Bonetto, 1975).

Bonetto and Wais (1990) consider the Paraguay to be an exception from the widely applicable paradigms of river ecology, by having its most extensive alluvional swamps in the upper instead of the lower reaches. In fact the Lower Paraguay is a rejuvenated outflow of a Glacial endorheic basin, which still functions partly as such. Even today, in the area of its outflow from the Pantanal, the Paraguay flows over shallow sills of as little as 0.80 m (Tricart, 1982) or 1.30 m (Valverde, 1972). The existence of the sills severely limits the navigation on the Middle and Upper Paraguay.

The hydrographic network of the left side tributary rivers is even more complex than that of the Paraguay. Some of these rivers are major rivers of their own. Rio Cuiabá, together with Rio São Lourenço have a combined medium flow of 652 m^3/s. Rio Taquari carries a medium flow of 299 m^3/s and the combined output of the rivers Miranda and Aquidauna is of 205 m^3/s (Oliveira Carvalho, 1986).

It is interesting to emphasize that the difference between maximum and minimum flow of the Paraguay is about tenfold, whereas the tributaries have seasonal flow differences amounting to 20 times (Rio Taquari) and 46 times (Rio Miranda). During the floods, the tributaries supply 2 to 5 times more water to the Paraguay than the own sources of this river.

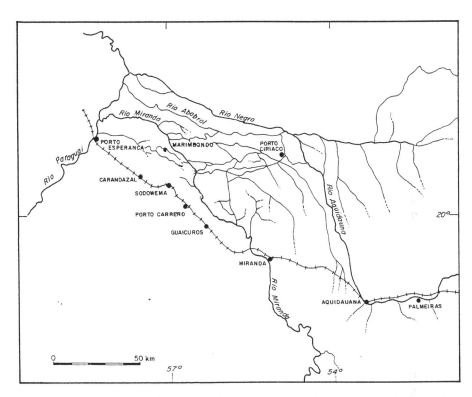

Fig. 18. The Rio Miranda-Rio Aquidauana-Rio Negro drainage system (redrawn from Adamoli, 1986). Note the track of the Railway.

In general, as one advances south, the left hand tributaries turn to be more and more torrential. A relatively small contributor, Rio Negro with a medium flow of 75 cubic m/s, has often minima of less than 1 cubic metre (Fernando Almeida Prado unpublished) and sometimes disappears altogether among the sandy alluvia.

Rio Cuiabá and Rio São Lourenço form a complex with the rivers Itiquira, Caracará, Casange and Bento Gomes and it is often difficult to decide in this network, which is the river and what name it should have. The outflows to the Rio Paraguay are deltaic and shifting and for instance today the mouth of Rio Taquari ("Taquari Novo"), is many kilometres distant from the old mouth of this river. It seems also that in historical times rivers São Lourenço an Taquari were confluent. The network of Rio Miranda and Rio Aquidauana appears as a hydrographic maze (Fig. 18).

By a rough calculation, the Paraguay exports only some 15–20% of the water supplied to the Pantanal. The rest percolates into the aquifer or evaporates. The summer flow in many rivers starts only when the water table is first saturated and reaches the surface.

On the Bolivian side, supply to the Paraguay is exclusively torrential. The rivers disappear during the dry season in the sediment, the so-called "sumidouros" and eventually become resurgent near the Paraguay or the great lakes. Besides the above mentioned case of Rio Negro, in the southernmost corner of the Brazilian Pantanal one can also find rivers which such names as Rio Perdido ("Lost River").

Perhaps it is adequate here to quote Ab'Saber (1988): "The Pantanal is the most complex intertropical alluvial plain of the planet and perhaps the least known area of the world in terms of correct alluvional knowledge".

8. The Amazonian connection

A much quoted but little known connection exists between the headwaters of the Jaurú, a major tributary of the Upper Paraguay and the headwaters of Rio Guaporé, itself a tributary of the Amazon basin (Grabert, 1967; Bonetto, 1975; Diegues 1990; Bonetto and Wais, 1990). it is evident that this connection is not a clearly defined one, like the connection through the Cassiquiare canal between the Rio Negro and the Orinoco, but it is remarkable how little we know about it. An important tributary of the Jaurú, Rio Aguapei, collects its waters from a swampy area which feeds also Rio Alegre, a tributary of the Guaporé. Vieira de Campos (1969) gives a very plastic description of these swamps in which the smallest trench or even a fallen log might change the direction of the drainage of a certain waterbody: either to flow to the Amazon or to the La Plata. Ab'Saber (1988) quotes Rosa and Santos (1982) when he says that the Guaporé depression is the link between the depressions oriented towards the south Amazon and those leading to the La Plata basin.

These "águas emendadas" ("joint waters") are not navigable to modern craft, but in the past they were extensively used for "portage" trade especially during the rainy season. Elisée Reclus (1895) even mentioned that in 1772 a channel was dug between the Rio Aguapei and Rio Alegre through which a six oars boat could pass. Because of little use, this waterway decayed. Adventurous canoe riders are reportedly using this interfluvial connection even today.

Without doubt, during more humid phases, the swampy contact between the two great river systems of South America through the "pantanals", must have been more intimate and less prohibitive to aquatic organisms. A number of medium-sized lakes in the area, such as Lagoa da Rebeca and Baía Grande may act as intermediaries between the two systems. As shown below, the contact was probably never large and deep enough in order to permit the colonization of the Pantanal by the giant Amazonian fish *Arapaima gigas* or by the large Amazonian aquatic mammals. These blurred headwater contacts deserve all the attention of the limnologists, no less than the famous Cassiquíare.

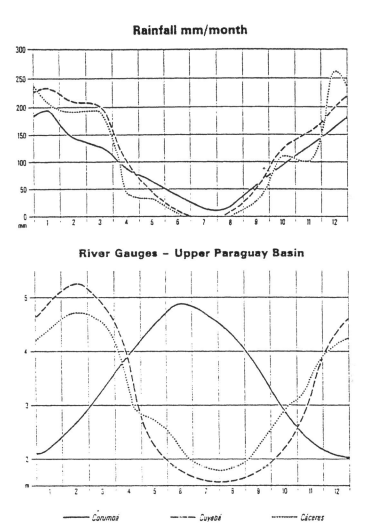

Fig. 19. Annual rainfall distribution and river gauges on River Paraguay (from Klammer, 1982).

9. The flood regime

Typically, the high waters in the Pantanal are out of phase with the rainy period (Fig. 19).

Comprehending and forecasting the floods in the Pantanal is and will be a very complicated task. Though maximum summer floods rarely exceed by much the 6 m, their occurrence, duration and amplitude is still a matter of conjecture, from place to place and every year. The economical importance of the problem resides in the fact that in the plains of the Pantanal, every

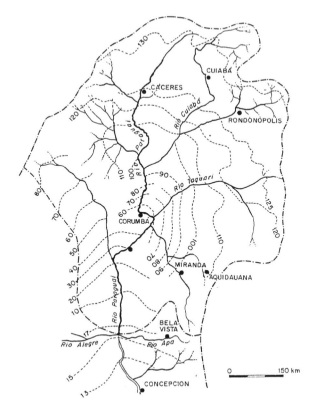

Fig. 20. Isorheocrones (equal high water levels) in the Upper Paraguay catchment (redrawn after Veloso, 1972). Note: figures indicate days before reaching on day 0, the confluence with Rio Paraná, further South. In the Pantanal the flood takes 120 days to reach the southern limit of the system.

additional metre of water cover results in the flooding of thousands of square kilometres more. Since the floods occur at different times in the different subregions of the Pantanal and with different force, their forecasting is perhaps one of the most investigated issues of the Pantanal (Valverde, 1972; Adamoli, 1986). The problem is, that besides the measurements on rio Paraguay at Ladário and Porto Murtinho which started in 1900 and on Rio Cuiabá which started in 1933, all the other data are spotty and based on personal information of the cattle ranchers and fishermen.

Several factors contribute to the complicated flood regime in the different parts of the Pantanal. First of all, there is the changing pluviosity of the headwater regions of the Pantanal streams. since these headwaters are situated in the tropical wet Amazonian climate as well as in the dryer climate of the Brazilian highlands, the floods do not coincide.

The water debt of the groundwater is different in the different subregions. In some areas the water table, 10 m deep has to raise before overflow starts.

Evaporative losses are also very high and variable, since the amount of water reaching the different baías and alagados differs as the streams change their flow direction.

The flood waters are spreading very slowly over the Pantanal, because of the minimal slopes. The floods after reaching the City of Cuiabá, take 17 days to reach the Paraguay. Floods of Rio Miranda take 15 days to reach the main collector. In the Paraguay itself, the "wave" of the flood travels very slowly through the lakes and the braids: it takes 3 to 4 months until the floods reach the outflow at Fecho dos Morros. Valverde (1972) presents data for a system of "isorheocrones", i.e., lines which link the points with equal flooding date (Fig. 20). The result is a phase difference of 120 days from the headwaters till the outflow from the Pantanal depression. Interestingly, from Valverde's map, it appears that the wave of the flood is retained at the southern outlet of the Pantanal, for 7 days.

The result is, that even in normal years, the floods are asynchronous in the Pantanal. As seen in the illustration by Klammer (1982) the flood in the Paraguay is 3–4 months delayed as compared to the Cuiabá (Fig. 19). The flood in the Paraguay may arrive, when the tributary rivers are already past their own high level period. Reverse flows and flooding from the collector to the tributaries may result. For instance on the lower Rio Miranda, two floods may occur: the first, owing to the high levels of the river itself, the second because of water reflux from the Paraguay.

For Rio Cuiabá, four hydrological seasons have been defined (Silva, 1990): 1. flooding ("enchente") between October and December; 2. High waters ("cheia"), between January and March; 3. Receding waters ("vazante"), between April and June; 4. Dry season ("estiagem"), between July and September. These seasons are different in the different parts of the Pantanal, out of phase sometimes with many weeks and even a few months.

The amplitude of the floods is different too. Annual floods are the rule along the Paraguay, Cuiabá and Miranda systems. The alluvional cone of the Taquari is flooded more seldom (Fig. 21).

Adamoli (undated report) classifies the floods according to their height. Around a medium value of 3.8 m and between the range of 3.5 to 4.5 m, are the ordinary floods. Exceptionally high floods can reach over 6 m and exceptionally low floods are below 3.5 m.

Adamoli (1986) gives the frame for the secular fluctuations of the flooding levels in the Pantanal. Accordingly, Rio Paraguay receives normally about 70% of its water volume from the Pantanal rivers and under normal conditions no water from the swamp areas reaches the collector river. That means that the whole scenario of the extreme flood conditions depends on the hydrological behaviour of the tributary rivers in each year and in every subregion and on the eventual overflow of the swamps.

There exists a certain cyclicity of high and low flood years.

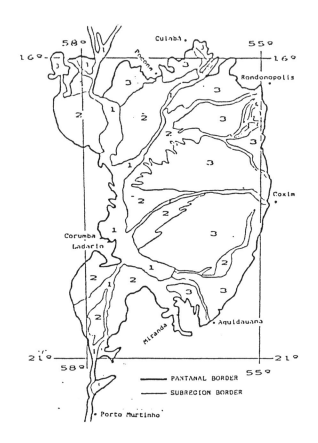

AREAS SUBJECTED TO FLOODING
IN THE PANTANAL

UNIT Nº	SURFACE km2	%	TYPE OF FLOOD	HEIGHT (m)	DURATION (months)	PREVAILING VEGETATION
1	21.803	16	Generalized	High	Long	Gallery Forest
2	46.863	34	Partial	Medium	Medium	Wet Grassland
3	69.334	50	Localized	Mild	Short	Savanna

Fig. 21. Subdivision of the Pantanal in accordance with the frequence of flooding (from Adamoli, undated report).

From the data gathered by Valverde (1972), Prance and Schaller (1982) and Adamoli (undated reports, 1986), it appears that extremely high floods were recorded in 1905, 1920, 1932 and 1954. After a very serious flood of

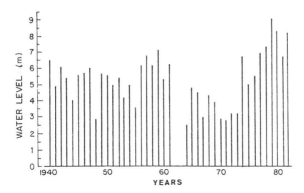

Fig. 22. Maximum annual flood gauges at Porto Murtinho (from Adamoli, 1986).

over 7 m in 1959, a low water period started which lasted for 14 years. A new series of high floods started in 1974, culminating with the record floods of over 9 m in 1979 and of 8 m in 1982 (Fig. 22). Since then, the local data indicate a new series of dry years and of low floods. 1994, with a flood level of a bare 2.5 m at Ladario, seems to be a culmination of this period of low floods. One can eventually speak of a periodicity of 12–15 years, of cycles of high and of low floods. This is presently the reigning paradigm, until a more detailed and reliable one will be found. In the area of Nhecolândia though, Mourão et al. (1988), report that 1974 has been a year of severe drought, contrary to the situation in Rio Paraguay. These authors prefer to speak of a periodicity of 5 to 7 years in the area of Nhecolândia.

Adamoli (1986) finds a good correlation between the high floods in the Pantanal and those in the Amazonian tributary rivers Araguaia and Tocantins. Junk (1993) goes further in suggesting a relation between the Pantanal flood cycles and the El Niño years. Following this line of thought, it would be interesting to check if correlation exists between the floods in Peru and Ecuador which are considered to be induced by ENSO events (El Niño Southern Oscillation) and the high floods in the Pantanal. Prediction of the flood regime is extremely important, because during low flood years agriculture and husbandry expands, whereas during high floods cattle mortality is high and fields are flooded. In the floods of 1974 for instance about 800,000 heads of cattle died, either drowned or of lack of pasture (Sucksdorff, 1989).

In the 70's a comprehensive international study has been carried out to further the understanding of the flood regime of the Pantanal, but the results were not as expected. Flood alarm system functions along Rio Cuiabá, where prevision depends mainly on the permanent gauging of the water levels of this river. Already in the floodplain, where other rivers and Rio Paraguay contribute to the floods, the alarm system is not too reliable. This is also the case for the Rio Miranda-Aquidauana system.

Fig. 23. A drying-out baía surrounded by scavanging birds of prey.

In the years of drought, like 1994 (Fig. 23), the Pantanal is ravaged by fire, mostly of natural cause and than the whole area turns into a "land of fire" (Sucksdorff, 1989) (see below). Though local and year-by-year prediction is already possible in several restricted and economically more important areas, long-time planning is still impossible in the Pantanal. The biota and also the humans have to live with the unpredictability of the environment. In a certain sense this is salutary for conservation.

10. The Pantanal lakes

More than anything else, the Pantanal is a land of lakes. There exists no census, but a prudent estimate would put the number of lakes in the tens of thousands. Most of them are small lakes of 500 to 1000 m diameter. On the lands of one farm alone (Fazenda Nhumirim), there are some 100 such small lakes (Mourão et al., 1988). The baías of the Pantanal are in their great majority morphologically well delimited, but they may increase much in size during the floods and during severe droughts, they may dry out completely. At high water stands they often coalesce into larger waterbodies of irregular branched shape (Fig. 42) and become part of the sluggish flow of the anastomosed waterbodies. When the waters recede, they separate and most of them regain the shape of rounded lakelets (Fig. 11). The classical areas of the round lakelets are on both sides of the Taquari, in Nhecolândia and Paiaguás, though they may be found also in other areas of the Pantanal.

In the area of Barão do Melgaço (Northern Pantanal), some baías were in 1985 six times larger than in the dry year 1967 (Silva, 1990).

The shapes and sizes of the lakes are often hidden beneath the dense carpet of the floating plants, the so called "batume". As the waters recede and salinity increases, the batume clears away.

Ab'Saber (1988) makes a rough classification of 4 lake types in the Pantanal, mainly following their sizes. There are first the large lakes along the Paraguay, traversed by the river or on its right bank. In addition there is the isolated large lake of Chocororé, near the escarpment in the northeast. A second category are the countless very young "oxbow" lakes in the floodplain triangle between the lower Cuiabá and the Paraguay. A third category are the above-mentioned round lakelets. Fourth, Ab'Saber mentions the karstic lakes on the Bolivian side of the river.

The large frontier lakes across the Paraguay are collectors of flood waters, eventual relics of much larger "pluvial" lakes. The largest of them, lakes Uberaba, Gaiba and Mandioré, receive also the overflow of the Cuiabá system. As a rule these lakes are interconnected by the Paraguay or communicate with the river and between themselves by channels, many of them artificially maintained. As a rule, they have also occasional inputs from the Chaco. For instance the medium-sized Lagoa de Cáceres which is the terminal lake of a Bolivian stream, is connected by a canal to the Paraguay and can therefore serve as the home port of the Bolivian navy. Canal Dom Pedro II which connects Uberaba with Gaiba, is 100 km long.

As already mentioned in the Introduction, there is almost no limnological knowledge about the large lakes. Many of them, like Lake Uberaba, are at time completely covered by vegetation. Their seasonal fluctuations are extreme: lake Gaiba covers 105 square kilometres at high water and shrinks to 55 square kilometres during the dry season. According to Tricart (1982), this is mainly due to the inflow of Rio Paraguay, but also to the high evaporation rate. Large areas of Gaiba are as shallow as 0.1–0.6 m.

Lake Mandioré exhibits similar features, fluctuating seasonally between 200 and 89 square km. These more northern lakes are typical overflow lakes. Lake Orion, fed by the Bolivian Coricha Grande is probably the most extreme case.

Lakes Cáceres, Lagoa Negra and Jacadigo are deeper lakes, probably of karstic origin, with less fluctuating levels and more reduced plant cover. information supplied by Reid and Moreno (1990) indicate that lake Jacadigo is a deep lake, without plant cover and unaffected by flooding from Rio Paraguay. Ab'Saber (1988) considers that eventually all the frontier lakes are of karstic origin.

During the very dry 1994, Lake Jacadigo dried out altogether, a phenomenon considered to be rare.

Nothing is known about the lakes of the Cuiabá-São Lourenço cone.

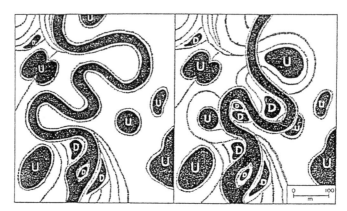

Fig. 24. *Wilhelmy's hypothesis of the genesis of the round lakes* (Wilhelmy, 1958). U = Meander lake ("Umlaufsee"); D = Embankment lake ("Dammufersee").

On the other hand, much data are on record from the round lakes of Nhecolândia. In aerial view, the multitude of these small lakes resembles in aerial view a field of water-filled bomb craters, as described by the war-experienced eyes of Wilhelmy (1958). The genesis of these regular shaped lakes is still a matter of much discussion. Wilhelmy (op. cit.) considered them to be "Umlaufseen", depressions surrounded by cut-off meander banks (Fig. 24). The banks of the short-cut meander, turned into cordilheiras. Wilhelmy further explains that these depressions receive ample groundwater supply through the sandy soils, because the shortcut river bed is situated at a higher level. Ab'Saber (1988) accepts this view.

Almeida and Lima (1956) saw the genesis of the round baías as the result of depressions caused by aeolian erosion. Klammer (1982) sides also with this aeolian hypothesis: accordingly, the cordilheiras which define the round lakes of Nhecolândia are old dunes left from the last arid episode. Klammer, however, concedes that this is only a partial explanation, since there are areas in the Pantanal which have such palaeo dunes without accompanying lakes. Here it is important to mention that the area of Nhecolândia is situated at 110 m, i.e., slightly higher than the surroundings and has a somewhat dryer climate.

Finally Tricart (1982) considers that the salt crystals which are formed in the dry littoral, form crumby granules with the dried mud, and these are easily blown away by the winds. An aeolian "sabkha" type of depression results. This process of deflation would have been probably much more active in past arid phases.

A third mechanism is being mentioned from time to time, to explain the genesis of the round lakelets: accordingly they would be karstic dolines. Eiten (1983) calls them "pseudokarstic" lakes. This type of origin can be probably dismissed, although karstic lakes are indeed found on the Precambrian calcareous soils of the west bank of the Paraguay (see above). More data about the round lakelets of Nhecolândia will be presented below.

11. Hydrochemistry

All the waters of the Pantanal known so far, are well-buffered neutral or alkaline waters. This is due in part to the mainly sandy sediments of the wetlands which are very permeable and well aerated, not permitting an acidification of the groundwater (Wilhelmy, 1958). Also the terrestrial vegetation does not seem to be of the type which supplies much humic substances to the environment. But still more important is the fact that the Pantanal is surrounded by calcareous, often karstic mountain formations.

The best example of a calcium carbonate-rich river is Rio Miranda. Draining the karst of Serra da Bodoquena, some of its headwater tributaries, like for instance Rio Formoso, are hypercalcic, actively travertine-depositing. Even in its lowland stretch, the Miranda presents high conductivity values. Mitamura et al. (1985) measured in this river conductivity values of around 128 μS/cm as compared with values around 40 μS/cm in Rio Paraguay and other rivers.

Rio Paraguay and Miranda had pH values between 6.82 and 7.55 (Mitamura et al., op. cit.), In Rio Cuiabá, Saijo et al. (1987) found lower pH values of 5.35 to 5.72. Near Porto Jofre on Rio Cuiabá, Silva and Pinto Silva (1989) measured pH between 6.0 and 7.3. In the smaller vazantes pH was between 6.0 and 8.0. In the alagados of the same area, pH values could be slightly acidic because of the decomposing vegetation. This phenomenon of "bad waters" ("águas ruins") is sufficiently widespread: After the first rains in the spring, the spill-off from the alluvional lands carries much decomposing vegetal matter as a consequence oxygen content is low, pH is slightly acidic and often fish death occurs. However, at the present stage of our knowledge, there are no permanently acidic "blackwaters" in the Pantanal system, even when here and there the name Baía Negra appears.

Goulding et al. (1988) reviewing the existing theories of blackwater genesis tend to accept the view that humic acids accumulate in soils with a very thin sandy cover. Deep sandy soils are not conducive to blackwaters. This is precisely the situation which exists all over the Pantanal.

The standing waterbodies of the Pantanal have a tendency to increased contents of dissolved solids, as is to be expected for a seasonally semi-arid climate. Saline lakes were well studied in the Nhecolândia area. These small and shallow rounded baías are rather bimodal, either completely fresh or saline (Mourão, 1989). Salinity data of these lakes are found in Cunha (1943), Brum and Sousa (1985) and Mourão (op. cit.). The salinas are typical "soda lakes" in which the predominant salt is natrium carbonate. They may reach pH values of around 10 and contain the nitrogen almost exclusively in the form of ammonia.

Mourão reported maximum conductivity values of 2,500 µS/cm. In July 1994 we recorded in a salina near the Embrapa farm in Nhecolândia a conductivity of 4,200 µS/cm (unpublished). Maximum recorded Na concentrations are 1,697 mg/l (Cunha, 1943) and 2,122 mg/l (Brum and Souza, 1985). The round salt lakes present all the characteristics of soda lakes, i.e., high productivity and a very restricted fauna.

The dry salt littoral and the completely dry salt pans are very much used as a mineral source for cattle.

There are no similar data from many other lake-studded areas of the Pantanal, neither are data available about the salinas of the Chaco periphery of the Pantanal.

A cumulative result of the hydrochemical properties of many Pantanal waterbodies, is the composition of the outflowing Rio Paraguay waters. As indicated by Bonetto (1975), the Paraguay presents a three times higher conductivity than its confluent, Rio Paraná. Whereas the waters of the Paraguay are characterized as being of the bicarbonate-chloride/sodium type, those of the Paraná are of the bicarbonate/calcium-magnesium type (Fig. 25).

In general, there are very few data about the nutrients and the chlorophyll values in the Pantanal waters. There exists only information from the lakes and streams on Rio Cuiabá and the streams associated with the lower Miranda. In lakes on the Transpantaneira near Porto Jofre, Saijo et al. (1987) reported total nitrogen values of 18.6 to 55.9 µg N/l and in the nearby Rio Cuiabá, 28.6 µg N/l. These authors emphasize the fact that in general ammonia represented more than 70% of the total Nitrogen. Nitrogen to Phosphate ratios were low, from 1.77 to 2.32, a fact which suggests according to Saijo et al. (op. cit.) a nitrogen limitation for algal growth. In Rio Miranda and some of its channels Mitamura et al. (1985) found total Nitrogen values of only around 7.5 µg N/l and Nitrogen to Phosphorus ratios of 0.8 to 1.1, again a very low ratio.

Heckman (1994) working in the flooded várzeas around Poconé, reports extremely low nutrient contents during the flood season. Nutrients increase dramatically during the dry season, in the small isolated baías and puddles. Heckman even speaks of oligotrophic conditions during the high waters and a rapid eutrophication during the dry winter.

Chlorophyll a content is extremely high in the baías and salinas of Nhecolândia studied by Mourão (1990): it went from 0.6 mg/m^3 in a "normal" baía to 2297 mg/m^3 in a salina. Silva and Pinto-Silva (1989) reported from some alagados, and corixos of Rio Cuiabá near Poconé, chlorophyll a values of 0.74 to 58 µg/l, whereas Saijo et al. (1986) found near Porto Jofre values between 1.82 and 8.66 mg/m^3. The comment of these authors is that the nutrient and organic matter contents of Rio Cuiabá are much higher than those of Rio Paraguay and of the southern rivers.

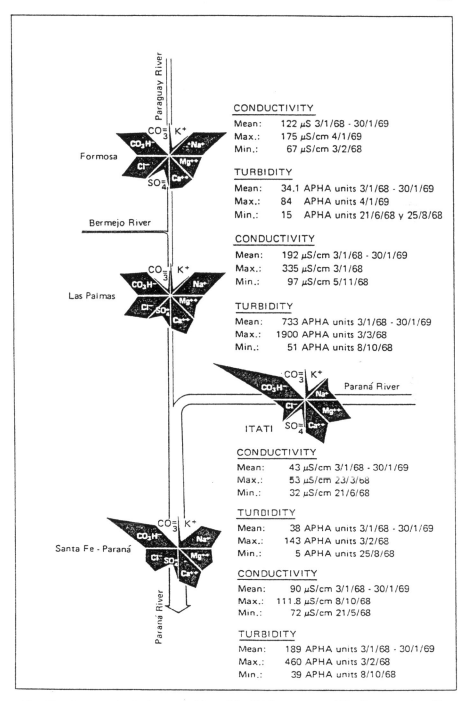

Fig. 25. *Comparative ionic composition of the Rio Paraguay and Rio Parana waters* (from Bonetto, 1975). Note: The most Northern graph and data are already some 400 km south of the Pantanal.

12. Soils

The Pantanal can be fairly clearly subdivided according to its soil types. This picture has been presented succinctly by Tundisi and Matsumura-Tundisi (1987) and in great detail by Pacheco do Amaral (1986). The determining factors, as usual in the field of pedology are a complex of granulometric, geological, geochemical, climatical, hydrological and vegetational factors.

Cardoso da Silva (1986) classifies the Pantanal into subareas according to the frequency of the floods (Fig. 21). Basically all the soils are hydromorphic soils.

From the viewpoint of the granulometry there are three clear subareas: 1. Sandy soils with less than 15% clay, mainly in the area of the Taquari cone, the triangle between the Paraguay and the Coricha Grande and the Nabileque panhandle; these are the relatively dry areas of the Pantanal. 2. Mixed soils, with more than 15% sand and less than 35% clay characterize the Cuiabá São Lourenço area. 3. Clay-like soils with more than 35% clay characterize the young alluvional areas accompanying the Paraguay, other rivers, and especially the Miranda-Aquidauana system (Pacheco do Amaral, 1986).

From a different point of view, the subdivision of the Pantanal is as follows: In the north laterites and planosols predominate, while in the south there is a predominance of more salty planosols and of solonetz. Gleis constitute the soils of the baías and the banhados.

Sodium content, resulting primarily from deficient flushing, increases from north to south. Low-sodium soils are located along the upper reaches of the Paraguay and the upstream areas of the Pantanal rivers. Sodium gradually increases along the lower reaches of the rivers and especially towards the Bolivian Chaco and the Nabileque panhandle.

According to Pacheco do Amaral (1986), more than 70% of the soils of the Pantanal present a low fertility. In the north this is due mainly to the protracted waterlogging. In the south the responsible is mainly the high sodium content. Based on all the considerations, Pacheco do Amaral (op. cit.) divided the Pantanal into 6 zones, according to the conditions for agrotechnical management.

13. How many Pantanals?

It is only natural that an area the size of the Pantanal needs to be subdivided. Administrative, traditional and objective criteria are, however, mixing in all these subdivisions. Since Correa Filho (1946) there is a widespread practice to speak of the "Pantanais Mato Grossenses", i.e., the "Mato Grosso pantanals".

Indeed in 1979 the state of Mato Grosso has been split in two. The frontier between Mato Grosso sensu stricto (MT) in the north and Mato Grosso do Sul (MS) follows the rivers São Lourenço and Cuiabá. Therefore one speaks

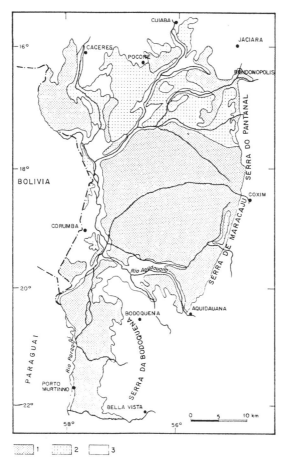

Fig. 26. Granulometry of the soils of the Pantanal (redrawn from Pacheco do Amaral Filho, 1986). 1 = sandy soils; 2 = medium texture soils; 3 = clay soils.

of a northern and a southern Pantanal. Adamoli (1980) divides the area into 10 pantanals (Fig. 27). Some of these are natural units and Adamoli gives phytosociological reasons for them. But the areas are too rigidly divided by the rivers in order to be natural. It is of course more precise to speak of the "Pantanal of Poconé" or the " Pantanal of Nabileque", but sometimes the definitions are not clear. For instance the Pantanal of Paiaguás is placed by Adamoli (op. cit.) to the south of the São Lourenço, whereas Ab'Saber (1988) places it to the north of this river. Of course, the subregions stop at the international border and subregions 6 and 2+3 are divided by the interstate border.

Natural sub-units have to be worked out, eventually along the geomorphological scheme of Klammer (1982) (Fig. 4) or the more practical considerations of Pacheco do Amaral (see above). Until such a scheme of natural

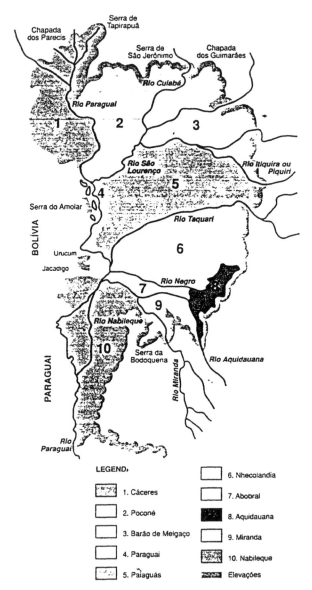

Fig. 27. The division of the Pantanal into different empirical subregions (from Magalhães, 1992).

subdivisions exists, it will be better to study the Pantanal as a mosaic of interacting areas. Useful is also a division of the Pantanal according to the occurrence and duration of the floods, into a High-, Middle-, and Low Pantanal (Adamoli, 1986) (Fig. 21).

The "Transpantaneira" a main permanently transitable road, which was to cross the Pantanal from Cuiabá to Corumbá, stopped near Porto Jofre when in 1979 the state of Mato Grosso was divided into two. This is perhaps good for reasons of preserving the environment, but can be seen as symptomatic for the dangers of artificial subdivision.

14. Regional connections

The Pantanal is the centre of a whole system of wetlands which are encountered in continental South America. The connections with the swamps of the Guaporé were already mentioned above. The wetlands situated to the north and the west of the Pantanal are even less known than the Pantanal itself. One fact has to be emphasized, however: these wetlands have more a character of seasonally flooded grasslands, perhaps with some perennial oxbow lakes ("curiches" in Bolivia). In the province Santa Cruz of Bolivia, the headwaters of Rio Otoquis, a tributary of the Paraguay meet with a distant headwater of the Amazonian Rio Mamoré, Rio Quimome, in a wetland area called "Banhados de Izozog" (Medem, 1983). The area is basically a dry Chaco, but in years of much rain, the whole system is a hydrographic continuum. In the Llanos de Moxos, on the Rio Mamoré in northern Bolivia, basically a flooded savanna, there are several large lakes, such as Lake Regagua and Lake Regaguado (Beck, 1984; Haase and Beck, 1989). According to Junk (1993) these Bolivian wetlands cover a surface of 150,000 square kilometres and contain over 1000 small lakes. Unlike the Pantanal, Llanos de Moxos is crossed by several blackwater rivers of the Amazonian type.

The floodlands of Rio Madeira, namely the flooded savannas of Humaitá in Brazil, were studied from the botanical aspects by Janssen (1986).

At the northeast, there is another area of extensive wetlands, namely the huge (20,000 square kilometre) river island of Bananal, formed by the rivers Araguaia and Tocantins (Junk, 1993). This is basically a large flooded savanna. It is worth mentioning, that the headwaters of Rio Cuiabá are separated by a very narrow gap on the Chapada dos Guimarães, from Rio das Mortes, a tributary of the Araguaia. According to some authors, the headwaters of Rio das Mortes and those of Rio Piquiri, a tributary of the Cuiabá, are joined by the lake Lagoa Agostinho (Medem, 1983).

Towards the south, already in Paraguay, Rio Paraguay is accompanied by a seasonally flooded Chaco. This has been classically studied by Carter and Beadle (1929) at Makhtlawaiya, some 60 miles west of Rio Paraguay. These authors studied the biologically important environmental parameters in the temporary pools and streams of that site.

Besides the high salinity values found in the residual pools of the streams, the importance of the huge temperatures (up to 40 °C) and of the lack of oxygen, on the biota were emphasized. Carter and Beadle (1929 a) call the tropical waters in which the oxygen regime is the crucial limiting factor, "brotocratistic" waters, and compare them to the hypolimnia of the stratified lakes.

More to the south, the wetlands accompany the Paraná in a very broad belt. These wetlands, extensively studied by Bonetto and his school are intimately linked to the Pantanal by the permanent supply of floating islands of the *Eichhornia* association and by many common species of fish and aquatic invertebrates.

The Pantanal therefore appears as a central hub in a network of large wetlands which connect the wetlands of the Amazon and those of the La Plata basin. During a considerable part of the history of the old Brazilian shield, there has been a permanence of wetlands in the centre of the continent. All these were interconnected at times. Therefore, even if the Pantanal in its present form, is a young formation, there is no doubt that it received very old freshwater biota which evolved in this intracontinental wetland system over long geological periods.

15. Introduction to the biology of the Pantanal

The Pantanal is a complex of aquatic and terrestrial environments in a permanent shifting interphase. It is difficult to make a clear separation between the biota of the dryland and the wetland and of the waters, as there is a considerable contingent of interphase dwellers and one can encounter such biota as marshland deer, lungfish, aquatic orchids and floating euphorbiaceans besides the classical aquatic and terrestrial biota.

With the exception of the isolated mountain islands of the morro's, all the terrestrial environments can be subject to flooding, either on a regular, seasonal basis, or exceptionally at high water levels. The rivers are shifting and the baías can occasionally dry out. Terrestrial plants have to cope both with conditions of high groundwater level and with extreme drought. Natural fires selected also plants with pyrophytic adaptations. Aquatic organisms face deficient oxygen conditions and very high summer temperatures in the standing waters and the slow flowing streams; on the other hand they have to be resistant to drought. Typical for the waters of the Pantanal are amphibious pulmonate snails, freshwater crabs and the above-mentioned lungfish of genus *Lepidosiren*. Practically all the plants are capable of amphibious survival.

Under such conditions, territoriality in animals is rare. Terrestrial mammals and birds follow the changing shorelines of the flooded areas, expanding when the swamps dry out. Water fowl adapt their roosting and nesting places to the changes in the distribution of the waterbodies. Fishes perform large-scale and long distance migrations, the so-called "piracemas" (see below). Floating plants accompany the slow streaming of the seasonal network of the vazantes and corixos.

Under the unpredictable conditions of the Pantanal extreme specialization, especially in food preferences, is impossible.

The Pantanal is sorely in need of intensive biological research. Good knowledge of the fauna exists only for the higher plants, the butterflies, fish and higher vertebrates. Already among the molluscs and the crustaceans there exist many white spots.

Brown (1986) discusses the reasons for this lack of knowledge. Most important among these reasons is of course the lack of access to many areas. There is also the imprevisibility of the conditions one will encounter on the sites. Because of this, Brown stresses the need to replace expeditionary studies with local, day-by-day Pantanal based research. In Brown's words "zoochronology" in the Pantanal is as important as zoogeography: Because of shifting conditions, one is bound to encounter in the same site and at similar dates completely different biota. Opportunistic organisms obliterate with their extremely high numbers the real diversity. Every year the image of the Pantanal is different.

The Pantanal is in a permanent state of successional changes. The sequence of terrestrial plant associations starts with the dry Chaco scrubland, but it does not reach the stage of a mature rainforest. In most cases it stops at the stage of a "pyroclimax" (Coutinho, 1990, see Fig. 31). in the most propitious setting, one encounters semi-deciduous forests or gallery forest. The dominant vegetation type of the dryer parts of the Pantanal is the "cerrado", the savanna vegetation.

The lakes of the region are too shallow and the climate is too windy to have real pelagic environments or a seasonally stable stratification in the waters. The rivers are all slow-flowing depositional streams. There are short episodes of torrential stream, the so-called "coivaras", feared by the fishermen, but they are too imprevisible to allow the development of a real hard-substrate stream fauna.

An obvious result of all this is the fact that the Pantanal is an area of very little endemism, especially among the terrestrial fauna and flora (Prance and Schaller, 1982; Brown, 1986). The present conditions are too fluctuating on a yearly and a secular basis, and they were probably even more so in the longer range of the Pleistocene climatic history. Perhaps with the exception of the isolated old hills, there has never been an environmental stability and a definition of isolated areas conducive to in situ speciation.

The cave fauna of the surrounding calcareous hills alone have yielded old endemics, as for instance the Speleogryphacea (Crustacea) found in a cave of the Serra da Bodoquena. As a rule, the flora and fauna of the Pantanal is composed of taxa which are found also beyond the limits of the system, expansive pioneer-type of species.

Brown (1986) considers the Pantanal to be first of all a biotic corridor and to a lesser extent a biogeographic barrier separating between the surrounding regions. In the taxa in which this aspect has been studied, the rate of endemic species does not exceed 5%. No student of any taxon ever proposed

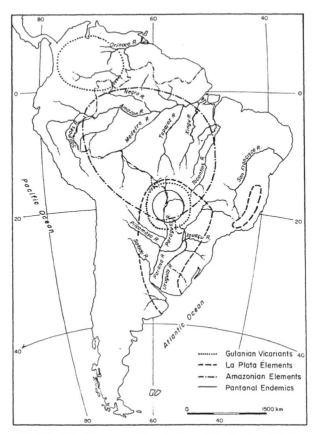

Fig. 28. The biogeographic relations of the Pantanal biota.

to separate the Pantanal as a biogeographic province of its own. Even for the aquatic zoogeography, the Pantanal belongs to the La Plata province and many putative endemics accompany the Rio Paraguay.

One cannot, however, exclude the assumption that for several typical aquatic biota such as the free floating plants, the caimans and the terapins, the Pantanal has functioned as a centre of speciation, being the major swampy waterbody of the continent. However, these taxa are today dispersed way beyond the limits of the Pantanal.

Perhaps the best way to characterize the Pantanal is to consider it as a biotic filter situated in the heart and the crossroads of South America. Only the most resistant species of the surrounding biogeographic provinces could adapt to the imprevisible environmental fluctuations of this region. However, those species could in exchange develop extremely big populations (Fig. 28).

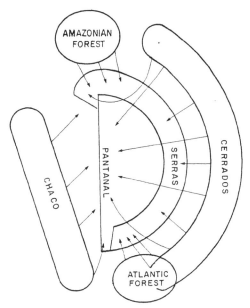

Fig. 29. Adamoli's scheme of the phytogeographic relations of the Pantanal (redrawn from Adamoli, 1986).

16. Phytogeography

The subdivision of the Pantanal among the different surrounding phytogeographic provinces is a subject very much discussed in the Brazilian literature, since Gonzaga de Campos (1912) and Hoehne (1923, 1936). Veloso (1972) speaks of a Chaco floristic subregion (1), a Cerrado floristic subregion (2) and a Transitional floristic subregion towards the Amazon forest (3). The most updated presentation of the subject is by Adamoli (1986). Accordingly, 4 distinct phytogeographic provinces divide between themselves the Pantanal (Fig. 29): 1. The Chaco province, which extends from Bolivia and Paraguay and is especially represented on the left bank of the Paraguay and in the southernmost part of the Pantanal. 2. The Cerrado province of the savannas and savanna woodlands of the Brazilian Planalto, which are the dominant province in the Pantanal. 3. The Peri-Amazonian province of semideciduous closed forest, best represented in the north but extending also in the gallery forests of the large rivers. 4. The Atlantic province, only recently added by Adamoli, is an extension of the southeastern forests of Brazil which appears in the Bodoquena massive but penetrates too into the gallery forests. Prance and Schaller (1982) add to this also a category of Neotropical plants with very wide distribution.

Pinto Paiva (1984) even tries to quantify the extension of the three main provinces in the Pantanal. According to him the cerrado province occupies

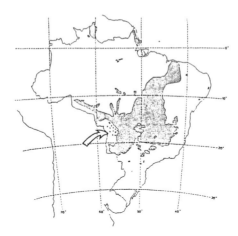

Fig. 30. General distribution of the cerrado in Brazil and the Pantanal seen as an archipelago of cerrado vegetation (from Coutinho, 1990).

97.338 square kilometres, the Amazonian province 29,213 square kilometres and the Chaco province 10,520 square kilometres.

According to Nogueira Neto (in lett.), the Pantanal can be considered an archipelago of "cerrado islands" in a sea of hydrophytic vegetation. This concept is expressed also in a map which accompanies Coutinho (1990) (Fig. 30). Adamoli (1986) considers that the phytogeographic provinces are so well delimited that the name "mosaic of Pantanal vegetation" is no longer justified. The facts in the field are, however, fairly complex. This appears clearly in Veloso's papers (1945, 1946, 1972), where dynamic successional aspects are considered. The polemics around this subject are also expressed by the Argentinean botanist Prado (1993), who considers that the forests of the calcareous Mato Grosso hills are not belonging to the Chaco province sensu strictu.

Veloso considers the vegetation of the Pantanal to be in a permanent successional process. The same dynamic aspect is emphasized also by Ab'Saber (1988) when he applies to the Pantanal vegetation the "Pleistocene refugium concept". The Chaco vegetation would be now in retreat, a mere relic of the vegetation which dominated the Pantanal during the cold-dry Pleistocene phases. The forests on the contrary were better represented during the Postglacial wet phase.

Today the cerrado predominates in the Pantanal and presents all its successional stages: from the "campos", a herbaceous savanna, typical for the sandy soils of the alluvional fans to the "cerradão", the climax formation of savanna forest which takes hold of the higher grounds of the hills. Using the thumb rule by Goodland (1971), the sequence is characterized as follows: 1. "Campo sujo" (unclean grassland), with trees up to 3 m height, 31 species

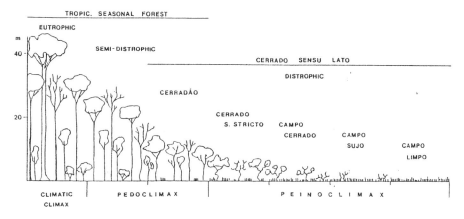

Fig. 31. A scheme of the cerrado succession (from Coutinho, 1990). The succession may reach several climax stages, imposed by soil fertility or by fire.

of trees and 60 species of herbs; 2. "Campo cerrado", with trees up to 4 m, 36 species of trees and 53 of herbs; 3. "Cerrado", with trees up to 6 m, 43 species of trees and 47 of herbs; 4. "Cerradão", with trees 9 m height, 53 species of trees and 24 species of herbs. This classification is rather artificial and forced, and has been corrected by Eiten, by Coutinho and others. Here I present Coutinho's (1990) graphic scheme of cerrado succession (Fig. 31).

The herbaceous form of the Chaco association dominates the clay alluvia of Rio Paraguay and the southern tributaries (Miranda, Apa and Nabileque). A Chaco type deciduous dry forest appears on the calcareous hills of the right bank of the Paraguay. Finally, the Amazonian and Atlantic forest elements appear in the bordering mountains, and mix in various proportions in the gallery forests.

It has to be mentioned, that in historical perspective, the Pantanal served probably as a main contact between the two rainforest types of Brazil: the Amazonian and the Atlantic one (Por, 1992). It is a "filtering bridge" which spans the "Great South-American Disjunction", i.e., the semi-arid Brazilian Planalto (Por, op.cit.).

There are very few endemics among the higher terrestrial plants. Prance and Schaller (1982) give a very extensive floral list of fazenda Acurizal, on the right bank of Rio Paraguay and on the shore of lake Gaiba. Among all the species, they mention only two endemics, the polygonacean *Coccoloba cujabensis*, growing on river banks and the loasacean *Mentzelia corumbaensis*.

The Pantanal is extremely rich in species of aquatic and semiaquatic plants. It is inhabited perhaps by the major concentration of aquatic plant diversity in the world (see below). As already mentioned, these are plants which by their very nature, have a large geographic distribution. Without doubt, some of these plants have originated in the largest swampland of South America. Recent studies have shown for example, that very closely related species of floating plants, so-called hydratophytes, like the genera *Salvinia* and *Eichhornia*, have a completely different biogeographical behaviour: some are local and endemic in South America while others became widespread global pests (Ashton and Mitchell, 1989).

17. Aquatic vegetation

* *Phytoplankton and algae*

Preciously little is known about the phytoplankton in the Pantanal waters. Useful data can be found again in the thesis of Mourão (1990) on the three lakes of Nhecolândia.

This author records only the generic names of the green algae. These are found in decreasing diversity in the freshwater Baía do Jacaré and the slightly saline Baía do Arame (see above). The green algae reported are *Botryococcus*, *Pediastrum duplex*, *Micractinium*, *Oocystis*, *Oedogonium* and *Spirogyra*. Mourão reports also 9 genera of Desmidiaceae from Baía do Jacaré, of which only 3 genera survive in Baía do Arame. Four species of Euglenaceae are found in the first lake, and two in the second. The Cyanobacteria do not appear in Baía do Jacaré, they are represented by 5 species in Baía do Arame and by 14 species in Salina do Meio, the soda lake. It is interesting that Baía do Arame is inhabited by typical freshwater bluegreens, such as *Microcystis aeruginosa* and *Oscillatoria limnetica*, which do not live in Salina do Meio. The largest contingent of bluegreens in the soda lake are species of *Oscillatoria*, three of them, according to Mourão, are synonymous with species of genus *Spirulina*.

As mentioned above, the Cyanobacteria of Salina do Meio and also in other salinas, are responsible for the extremely high values of primary production in these waterbodies.

Lamonica Freire et al. (1992) studied the pigmented Euglenophyta near Poconé and reported two genera, *Lepocinclis* and *Strombomonas*, which were not reported in Mourão's (1990) work.

Heckman et al. (1993), studied the cosmopolitan algae in ephemeral waterbodies of the Pantanal. it results that during the flood season algae are almost absent in the plankton, presumably because of nutrient limitation (see above). In the isolated pools left during the dry winter, algal blooms appear, formed especially of species of *Scenedesmus*. Before the pools dry out and water temperatures can be in excess of 40 °C, red carpets of a bloom of *Euglena sanguinea* are widespread. Heckman et al. (1993) report a total of 23 species of Chlorophycea, 14 species of Euglenophyta and only 3 species of Cyanobacteria.

Fig. 32. Floating plant cover of a baía, mainly of Salvinia.

* Submerse rooted vegetation

Submerse rooted vegetation practically does not exist in the Pantanal. Probably the shading from the floating vegetation is an extreme limiting factor. Only in the more saline baías, like Baía do Arame in Nhecolândia (Mourão, 1990), where the cover of floating plants is repressed, the bottom is covered by a dense growth of Characea.

* Floating vegetation

The Pantanal is inhabited perhaps by the highest diversity floating aquatic plants in the world. This is due both to the fact that some species are endemic or originated in the area and to the diversity of the lentic environments (see table 1 on p. 51). The carpet-like cover of floating plants (Fig. 32) probably competes or even exceeds in extension, that of the graminacean and other grasses. They represent without doubt the major vegetal biomass and the most important primary producers in the Pantanal. They are either rooted floating plants or free floating plants. In the terminology of the botanists, they are "hydatophytes". By dividing between them the huge surfaces of the baías and slow flowing corixos, the floating leaves, both escape the lack of oxygen in the warm shallow waters and contribute through dense shading to the oxygen dearth of these waters.

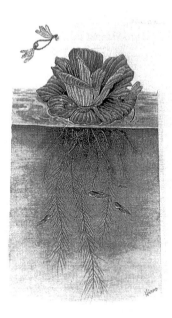

Fig. 33. Eichhornia crassipes and Pistia stratiotes (from drawings by Umiko, São Paulo).

There is a whole sequence from plants which are rooted in the bottom and expose their floating leaves at the surface, like *Victoria*, through plants which are rooted but carry dense tufts of aquatic roots, to free-floating plants.

Among the floating rooted plants I included also the peculiar aquatic orchid, *Harbenia aricaensis* which exposes its flowers above the surface of the water, being rooted often more than one metre deep. Among the free-floating plants the species of *Utricularia* are an exception: they float entirely submersed, below the carpet of the other free-floating plants. The "carnivorous" supplementary feeding of these plants possibly compensates for the reduced possibility to perform photosynthesis in this shaded environment.

Aquatic roots are a necessary adaptation in the warm and often oxygen deficient waters of the Pantanal baías. Among the Pontederiaceae there are species which are mechanically rooted but provided with masses of aquatic roots and from among them *Eichhornia crassipes* ("aguapé") became the prototype of a free floating plant.

According to Ashton and Mitchell (1989), many of the hydatophytes, and especially the free floating ones have advanced capacity to reproduce vegetatively, through active fragmentation. The free floating species represent an "active" way of life and it is not the result of accidental detachment from a rooted stand.

Fig. 34. The aquatic orchid Harbenia arikaensis (from Sucksdorff, 1987).

The taxonomy of the free floating plants is complicated by the fact that they present different phenotypes. As shown by Ashton and Mitchell (1989), *Eichhornia crassipes* presents a "colonizing" form which is free and a form with a completely different habitus, the "mat form", which is part of a floating island. In genus *Salvinia* ("orelha de onça"), there are 4 species in the area of Brazil and Argentina, each characterized by a different chromosome number and different ecological and reproductive behaviour. They too may have different "colonizing", "survival" and "mat" phenotypes (Ashton and Mitchell, 1989). The competitive colonization of the water surface by the different hydatophytes, is a very interesting subject. There exists a certain horizontal zonation between the species, as shown by Vali Pott (in lett.).

There are also differences in the environments settled: for instance *Reussia subovata* inhabits preferentially the flooded brejos, whereas *R. rotundifolia* is frequent in rivers and corixos. There are certain areas of the Pantanal in which *Pistia* ("erva de Santa Luzia" or "alface de água") does not appear, while in others it occupies a dominant position.

Poi de Neiff and Neiff (1984) working on the Lower Paraguay and the Paraná, describe a "micro-succession" in the floating vegetation of lakes, in which *Eichhornia crassipes* is a sort of "climax" which starts with smaller forms like *Salvinia, Azolla*, etc.

Da Silva and Pinto-Silva (1989) and Da Silva (1990) studied the productivity of the hydrophytes of the northern Pantanal, emphasizing local and seasonal differences. According to Neiff (1990), *Eichhornia crassipes* presents the highest productivity values for the whole La Plata basin.

Moreover, this high productivity is not dependent of the hydrometric fluctuations, whereas other floating plants, especially the rooted ones present high growth and productivity mainly during the rapid rise of the flood level. The majority of the organic matter produced by the aguapé reaches the rest of the ecosystem indirectly, through decomposition of the dead plant mass (Neiff, 1990). The *E. crassipes* populations cannot support drying out of the waterbodies in which they are "captured". The resulting layer of dead organic matter, the "batume" (Silva and Silva, 1992), is afterwards colonized by a shrubby "piuná", composed among others by weedy and noxious species of *Ipomoea*.

The aquatic roots are without doubt the environment of the Pantanal waters which is richest in invertebrates. Rich periphyton and vegetal detritus captured among the many fine roots, supplies a very rich food base for the invertebrates.

Without doubt, the recycling of the production of the floating plants starts already among their roots, with the plant still alive. This process has, however, not been quantified till now. (see below).

At the surface of the leaves, there is also a specific fauna. One link of it, for example, are typical semiaquatic Orthoptera, some of them show some degree of specificity (Moraes Paula and Aguiar Ferreira, 1994): *Cornops aquaticum* on the Pontederiaceae *Paulinia acuminata* on *Salvinia* and on *Pistia* and *Marelia remipes* mainly on Nymphaeaceae.

Heckman (1994) adds to this list also the curculionid beetle *Neochetina eichhorniae* and the argiopid spider *Actinosoma pentacanthum*.

At the other end of this "epipleustonic" food chain one finds *Jacana jacana* ("jaçanã") which walks and feeds on the floating leaves and even builds its nests on them. Some species of galinules live on the rafts of *Eichhornia* and several species of frogs and even a species of caiman (see below).

Fig. 35. A small camalote on Rio Miranda.

* The Camalotes (Fig. 35)

A complex of three free-floating plants, namely *Salvinia, Eichhornia crassipes* and *Pistia*, establish floating islands called "camalotes", sometimes many metres in diameter, which become detached from the plant cover of the baías and follow a many months long journey downstream Rio Paraguay, until through the Paraná and La Plata, they reach the ocean. Sometimes, the rivers are crowded with these stately floating islands. The camalotes are also called "embalsados" (Argentina) and "batumes" (Paraguay).

Expressing the rhythm of droughts and floods, the camalotes may accumulate, coalesce, and be carried away afterwards and sometimes fragmented (Bonetto, 1975).

Sometimes the islands can be retained temporarily by being "captured" by rooted hydrophytes such as *Paspalum repens* and *Ludwigia* (Dioni, 1967). The larger camalotes can be secondarily colonized by rooted emergent plants. Prance and Schaller (1982) count among these adventive plants *Cyperus haspan, Scirpus cubensis, Rhynchospora corymbosa* and *Elocharis* sp. Even the arbust *Senna pendula* may grow on camalotes.

The camalotes represent self-contained small ecosystems, in which dead radicular tissue is recycled, together with other captured detritus, by a "periphyton" of bacteria, fungi and algae, which serve as food for a rich invertebrate fauna and some predators, especially fish. The aquatic roots thus can absorb directly recycled minerals from the "rhizosphere" of their roots (Dioni, 1967; Por and Rocha, in press).

On the other hand, as shown by Jedicke et al. (1989) and Heckman (1994), surplus oxygen from the aerenchyme of the floating plants, as well as some photosynthetic oxygen, form a film of around the roots and rootlets, i.e., a microenvironment which is much richer in oxygen than the surrounding waters. The presence of oxygen around the roots attracts especially in the warm summer, several species of fish to live among the roots. These serve also as a hiding place and a rich feeding substrate for the young fingerlings (Sazima, 1988; Jedicke et al., 1989). However, as shown by Junk (1973), below the large camalote islands, the amount of oxygen decreases as one progresses towards the centre of the camalote, where decomposition processes are quantitatively more important.

Although there seems to be no difference in the invertebrate fauna living among the roots of the three main camalote-forming plants, it seems that there are successional changes, as the floating mats descend downstream. This aspect as well as the whole problem of the time the descent takes as well as the amount of hydatophyte biomass which leaves the Pantanal merit further study.

The camalotes carry with them also a significant terrestrial macrofauna. Caimans travel downstream and reach Argentina and there are also unconfirmed reports of jaguars which are carried away on the islands.

Camalotes are known also in other tropical continents. In the world literature they are called "sudd", "kirtas" or "floatants" (John, 1986). They gave their name to the Sudd swamps of the White Nile. They are essentially composed of the same plants, namely *Eichhornia crassipes*, ("water hyacinth"), *Salvinia molesta* and *Pistia stratiotes* ("Nile cabage", "soldier weed"). At least the first two of these species are pests which spread and invaded coming from the swamps of South America. The circumtropical distribution of *Pistia* is probably of older origin.

18. Terrestrial vegetation

As already mentioned above, the phytosociology of the Pantanal is well studied, although, as usual in such cases, there exist different opinions and classifications. I shall try to give an overview of this subject, using my liberty as a non-specialist. In order to simplify a very complex phytosociological picture, as presented by Veloso (1972) or by Valverde (1972), I shall largely rely on Prance and Schaller (1982) and on Dubs (1992). It is worth mentioning, that the Pantanal vegetation is often characterized by monospecific tree stands, called "parques". Typical parques are the "paratudais" of *Tabebuia caraiba*, the " carandazais" of the palm *Copernicia alba* (Fig. 36), "acurizais" of the palms *Attaleia* spp., and "buritizais", of the palm *Mauritia vinifera*. Among these dominant palm species, the first is frequent in the southern areas, while the "buriti" is encountered at the northern edge of the Pantanal.

TABLE 1

The floating hydrophytes of the Pantanal. This list was prepared with the help of Dr. Ilana Herrnstadt, Jerusalem (free-floating plants marked with *).

Pterophyta
    Azolla filicoides *
    Salvinia auriculata *
    Marsilea polycarpa
    Ceratopteris pteridoides *
Anthophyta
Gramineae
    Paspalum repens
Araceae
    Pistia stratiotes (Fig. 33) *
Euphorbiaceae
    Phyllanthus fluitans *
Alismataceae
    Echinodorus tenellus
    Echinodorus paniculatus
    Echinodorus uruguaiensis
    Echinodorus brasiliensis
Cabombaceae
    Cabomba australis
    Cabomba piauhyensis
Hydrocharitaceae
    Limnobium stoloniferum
    Elodea (=Anacharis) ernstae
Hydrophyllaceae
    Hydrolea spinosa
Limnocharitaceae
    Hydrocleys nymphoides
    Hydrocleys modesta
    Limnocharis flava
Onagraceae
    Ludwigia (=Jussiea) sedioides
    Ludwigia natans
    Ludwigia peruviana
    Ludwigia nervosa
    Ludwigia torulosa

Table 1 (continued)

Nymphaeaceae	
Nymphaea amazonum	
Victoria cruziana	
Menyanthaceae	
Nymphoides indica	
Nymphoides humboldtiana	
Lemnaceaea	
Lemna sp.	
Pontederiaceae	
Pontederia lanceolata	
Reussia subovata	
Reussia rotundifolia	
Eichhornia azurea	
Eichhornia crassipes (Fig. 33)	*
Heteranthera limnosa	
Leguminosae	
Aeschynomene fluminensis	
Lentibulariaceae	
Utricularia foliosa	*
Utricularia poconensis	*
Orchidaceae	
Harbenia aricaensis (Fig. 34)	

* *Amphibious herbaceous vegetation*

This association grows under permanently wet, the temporarily submerse conditions of the alagados and brejos as well as in shallow baías. These are first of all the graminaceans *Cyperus giganteus* and *Scirpus validus* ("pirí") as well as *Paspalum repens* ("capim d'agua"), the reed *Typha dominguensis* ("taboa"), the maranthacean *Thalia geniculata*, *Polygonum hispidum* ("fumo bravo") and other species of *Paspalum*.

* *Herbaceous campo vegetation*

In relatively few cases does the cerrado vegetation develop in the Pantanal into its woody seral associations (Eiten, 1983). Most of the area is occupied either by herbaceous floodland vegetation, with sometimes a predomination of arbusts on dryer soils or after major disturbances.

The campos, or campos limpos, dry or seasonally flooded grasslands are primarily inhabited by Graminaceans ("capim") such as *Paratheria prosta-*

Fig. 36. Carandazal: stand ("parque") of Copernicia alba (from Dubs, 1992).

ta ("capim mimoso"), the best fodder grass, *Panicum maximum* ("capim colonião") and other species of *Panicum*, *Oryza subulata* ("arroz bravo"), *Setaria geniculata* ("capim mimoso vermelho"), *Portulaca* spp. ("onze horas"), *Arachis glabrata* ("amendoim do campo") and others. Killeen and Hinz (1992) analyzing the distribution of 113 species of Graminacea in eastern Bolivia, reach the conclusion that the complex of the Pantanal, because of its varied microtopography, has the richest grass flora of all vegetation types studied by them.

Some herbaceous plants and small arbusts are distasteful or even venomous to cattle, such as *Ipomoea fistulosa* ("algodao bravo"), *Solanum malacoxylon* ("espichadeira") and other species of *Solanum* ("fruta do lobo", "jurubeba") and *Vernonia* spp. ("assa peixe").

After much overburning or excessive pasture, some arbusts become important in the campos, like *Brysonima intermedia* ("canjiquera"), *Erythroxyllum tuberosum* ("mercúrio bravo") and the above-mentioned *Vernonia*. Also an invasive graminacean *Aristida pallens* ("barba de bode") can expand (Valverde, 1972). Veloso (1972) mentions in this connection also *A. capillacea* ("capim carona").

On higher and dry grounds, the cattle ranchers introduced exotic herbs, such as *Brachiaria*. On some of the huge ranches of the banks and of the

Fig. 37. A flooded "murundu" field, wooded termite mounds (from Dubs, 1992).

large agro-industrial conglomerates, the seeding of these foreign pasture is done by air.

* Gallery forests

Gallery forests, called in Brazilian "matas ciliares" or "saranzais" cover the higher banks of the major rivers. Among the most typical river bank trees are *Triplaris formicosa* ("pau de formiga", a highly myrmecophilous tree), *Rheedia brasiliensis* (" bacuparí do rio"), different species of *Inga*, species of *Vochysia*, species of *Ficus* and especially the "mata pau" the strangling ficuses. The gallery forests contain rainforest trees like *Pterocarpus rohri* and *Pithecolobium multiflorum*, ("canafístula"), *Guarea macrophylla* ("caiarana") and *Licania* spp.

Ficus elliotiana is an endemic species of the Pantanal, specially adapted to undergo and survive periods of drought. (Neves et al., 1993).

* Murundu islands (Fig. 37)

The inundated cerrado fields are dotted by little elevations of up to 1–2 m, so-called "murundus" on which grow two species of *Tabebuia*, namely *T. aurea* ("ipê amarelo") and *T. caraiba* ("paratudo"), or more rarely *Vochysia rufa* ("cambará"). These trees, typical for dry cerrado, take profit of the building activity of the termites, such as for example *Rotunditermes bragantinus* (Dubs, 1992), or of some ants of the leaf-cutter Attini.

This monospecific formations are well known in literature and already Carter and Beadle (1929 a) mention them in the Paraguayan Chaco. Eiten (1983) considered them to be very frequent in the wet cerrados covering sometimes several tens of hectares and at a density of 80% cover (Eiten, 1983). The mounds protect the trees and their roots from waterlogging and eventually also from the brush fires of the dry season. The murundus crowned with their tree form a miniaturesque archipelago of equally spaced islands in the flooded campos. The equal distance between the islands is explained by Junk (1993), as expressing the territorial limits between the termite colonies. Sometimes such paratudais cover surfaces of many square kilometres.

The interaction between the specific tree species and the termites has not been studied in detail. In many cases though, the trees grow on deserted termite mounds. Valverde (1972) suggests a symbiotic relationship between the termites which create a watertight medium in their mounds for the roots of the *Tabebuia* tree, whereas the trees supply the termites with wood in an environment in which otherwise there would be no tree. This view is discussed by Walter and Breckle (1984) who tend to think that such "Termitensavannen", at least the African ones, are based simply on the topographic advantage a dead termite nest can offer to the trees.

A more detailed study of the reduced number of cerrado forest trees which can enter the "murundu" association, is by Oliveira Filho (1992).

Dubs (1993) emphasizes the role of these small islands of cerrado vegetation, as refugia for the terrestrial biota during the periods and years of flooding.

* *The cerrado*

This is a savanna vegetation, ranging successively from dry herbaceous fields ("campos limpos"), through shrub savanna ("cerrado") and savanna forest ("cerradão") and eventually a semideciduous forest (Goodland, 1971; Eiten, 1983; Coutinho, 1990). As already mentioned, the "terra typica" of the cerrados is the Brazilian Planalto.

Among the many species of trees which compose the cerrado forests, the most widespread ones in the Pantanal are *Caryocar brasiliense* ("piqui"), *Qualea grandiflora* and *Q. parviflora* ("pau-terra"), *Kielmeyera coriacea* ("pau santo"), *Curatella americana* ("lixeira"), *Stryphnodendrum obovatum* ("barbatimão"), *Hymenaea stigonocarpa* ("jatobá"), *Pterogyne nitens* ("balsamo"), *Vochysia tucanorum* ("pau de tucano"), *Vitex cymosa* ("tarumá"), *Diptychandra glabra* and *Tabebuia caraiba* ("paratudo"). Often cerrado type forest covers also seasonally flooded floor. *Copernicia alba* forms monospecific palm stands in the flooded areas of the southern Pantanal.

Prance and Schaller (1982) report diversities of 22 to 36 species of trees on a hectare of well developed cerradãoes.

* *Semideciduous and deciduous forests*

These are found on hilly and slope areas, in the north with many Amazonian elements, such as *Hymenaea courbarii*, ("jatobá') in the west with more xeric species of the Chaco forest type. Among the many tree species, more typical are *Enterolobium contortisiliquum* ("orelha de pau"), *Piptadenia macrocarpa* ("angico"), *Combretum leprosum*, species of *Copaifera* ("pau d'oleo") as well as several palms. On dryer slopes, these forests are called "quebracho", and contain many Chaco species, such as *Schinopsis* ("quebracho colorado"), *Aspidosperma* cf. *ulei* ("quebracho blanco") and *Caesalpinea paraguariensis*.

On a 1 ha plot of semideciduous forest of Fazenda Acurizal, Prance and Schaller (1982) counted 35 species of trees with a trunk diameter of over 15 cm.

On dry slopes, xerophytic elements appear among this forest, such as the "barriguda" *Ceiba pentandra* and *Cavallinesia* which accumulate water in their trunks, somewhat baobab-like.

Many spinous Leguminosae, such as *Acacia*, *Mimosa* and *Prosopis* are found in these associations, jointly with the "quebracho", *Schinopsis* (Eiten, 1983).

Also specimens of cactaceans, such as the "mandacuru" (*Cereus bonplandii* and *Pereskia saccharosa*), are common.

Several arboreal species of these dryland forests are encountered also in the dry caatingas of northeastern Brazil. For instance, the typical caatinga species *Zyziphus joazeiro* ("juazeiro") is frequent on the calcareous slopes of the western Pantanal edge.

It is important to mention that the two endemic plants reported by Prance and Schaller (1982), namely *Coccolobium cujabensis* and *Mentzelia corumbaense* are part of this type of forests.

The above-mentioned acuri palms are also part of this type of western Chaco forests. On the dry slopes one finds also an understorey of bromeliads of genus *Bromelia* ("caraguatá") and *Dyckia* ("croatá").

* *The flora of the monadnocks*

Strewn over the whole extent of the western and southwestern Pantanal, there are tens of small and steep Precambrian calcareous hills, "morros" or "inselbergues" (from the German "Inselberg"). Their base seldom surpasses 1000 m in diameter and they reach up to 100–200 m. Some of these morros are so small that they do not appear on the 1/1,000,000 maps of the RADAMBrasil project. They stand above the swamps as real mountain islands, with their bases buried in hundreds of metres of sediment thickness. Because of their steepness, they generally escaped felling, farming and pasturing.

The classical morro is Morro do Azeite on the lower Rio Miranda (Fig. 14). It presents from the bottom to the top a very clear zonation, which starts with a gallery forest, continues with a cerradão and peaks out in a dry forest with *Cereus* and *Ceiba pentandra* and sometimes with a typical "quebracho" shrub (see above).

The sequence of the morro flora represents an ideal actual model to study the fluctuations which the vegetation of the Pantanal suffered in the Pleistocene. For instance the "Chaco" species are evident relics from a dryer period, when this type of vegetation prevailed in the Pantanal.

Like their namesakes, the water-surrounded islands, the morro's are also ideal places to study small-scale evolutionary changes due to isolation. In brief, they are perhaps among the most important sites for preservation in the Pantanal.

19. Fire in the Pantanal

As mentioned above, the Pantanal has been called "Land of Fire" by Sucksdorff (1989). This name can be applied to all the cerrado lands. As shown in the last of his many studies, Coutinho (1990) considers the periodical burning of the cerrado to be a natural and even necessary environmental action. Natural fires occur towards the end of the dry season, i.e., August-September and the plants and animals are fire-adapted. The woody plants are pyrophytes, with long woody underground organs, the herbaceous vegetations sprouts after the mineral enrichment caused by the fire.

Many shrubs and trees of the cerrado produce flowers and seeds in the period which follows the burning-over. The animals have their own "pyrozoic" adaptation. According to Coutinho (1990) several mammals and birds are grey-coloured in order to mix mimetically with the background of burned trunks. Many animals are adapted "fire followers": the most classical case is that of the *Buteo albicaudatus*, "gavião fumaça" or the "smoke hawk" (see also below). A good reason for the extremely high density of cerrado animals in the Pantanal, is the fact that they find easy shelter during the fires in the wet and water-covered areas which surround the cerrado islands.

Coutinho (1990) reports that the pantaneiros use fire as the most primitive, but also very effective way of replenishing their pastures with grass vegetation. Burning simply provides large amounts of mineral ash for massive herbaceous growth. If this practice is done out of the season it is dangerous, because it does not replicate any more a propitious natural cycle to which the whole system is well adapted. Coutinho (op. cit.) also emphasizes the need for using controlled burning-off in the management of the cerrado reserves, in order to avoid invasion by alien plants.

Such managed burning is already successfully practised in grassland reserves in North America.

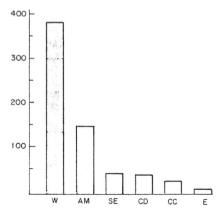

Fig. 38. Biogeographical analysis of the avifauna of the Pantanal (from Brown, 1986, simplified). E = endemics; CC = Chaco species; CD = Cerrado and Planalto species; SE = Southeastern Atlantic species; AM = Amazonian species; W = species with wider distribution.

20. Zoogeography

Brown (1986) summarized his firsthand knowledge about the distribution of 657 species of birds and 498 species of lepidopterans reported from the Pantanal. He used also second hand data about the amphibians, reptiles and mammals. Dubs (1992) exposed succinctly Brown's biogeographic conclusions about the avifauna of the Pantanal. Less than 2% of the species are endemic. For about 42% of the species, the Pantanal is a corridor from one bordering province to another. Of these, 25% are open landscape cerrado- and Chaco species and 20% are forest species which transit from the Amazonian to the Atlantic rainforests. For 40% of the species, the Pantanal is a faunal barrier, i.e., they reach their distributional limit within the region. Of these, 23% species are of Amazonian origin, mainly confined in the north of the Pantanal, 6% are Atlantic species, mainly found in the south and 6% are cerrado species. The remaining 12% of the avifauna are aquatic birds of wide distribution (Fig. 38).

Most important is the fact that the Pantanal represents a barrier to the Amazonian species. This is due probably to the effect of the "friagens", the occasional cold spells which reach the Pantanal from the south (Willis, 1976). An important factor is the impoverishment in the stock of the Amazonian rainforest trees, with which many birds are connected specifically.

From among the lepidopterans Brown (1986), analyzed the Papilionidea and the Hesperidae. The results are very similar to that of the birds. There are many Amazonian species which encounter their limits in the northern Pantanal. Among the Nymphalidae, the rate of endemism of nearly 5%, but mainly at the infraspecific level.

There are no endemic species among the mammals. The majority of the mammalian species are typical for the cerrado and their connections are more with the Atlantic forest fauna. Many Amazonian bats reach only the northern areas. The Amazonian manatee *Trichechus* and the dolphins did not reach the area. Brown (1986) remarks that among the aquatic herpetofauna there is a higher rate of endemism than in the birds. The endemic genus *Dracaena*, an aquatic teid lizard is a good example.

Again on the negative side, Brown (op. cit.) mentions the lack of salamanders and of pipid frogs from the Pantanal, species very well adapted to standing waters elsewhere.

There are 405 species of fishes on record from the Pantanal (Marins et al., 1981). This is a considerable diversity, but of course no match to the Amazonian one. The giant of the Amazonian fishes, the pirarucu *Arapaima gigas* does not inhabit the Pantanal. From Banarescu's (1990) study, it results that the majority of the fishes are of Amazonian origin, but there is also a considerable contingent of taxa pertinent to the La Plata basin. Pinto Paiva (1984) emphasizes the predominance of the lenitic fish species in the Pantanal. All the species, which in the Amazon basin are connected with swift flowing, lotic river conditions are absent or very much restricted.

Among the fish, there is a fair number of endemics. The freshwater flounder genus *Hypoclinemus* is endemic to the Paraguay basin. Also there are three endemic species of the freshwater stingray *Paratrygon* on record. These endemics might plead for the existence of older than the Pliocene waterbodies in the Pantanal. They are real palaeoendemics.

Among the characinids, there are several cases reported by Banarescu (1990). The extremely common *Brycon orbignianus* "piratputanga", is a Pantanal endemic. Among the Serrasalmidae, the family of the "piranhas" there are several species endemic to the Upper Paraguay, while other species are present in the La Plata system as a whole. Also among the cichlids there are several endemics.

Banarescu (op.cit.) presents also several interesting cases of vicariant distributions between the Paraguay basin and the northern basins of the Orinoco and the Magdalena, with no intermediate forms in between, i.e., in Amazonia. Such is the case of the characinids *Xenurobrycon* and its northern vicariant *Markiana*, respectively, of *Phoxinopsis* and *Grundulus*.

A similar picture is found also in the loricarid genus *Xiliphus*, which has one species in the Paraguay basin and two others in the Magdalena basin. In all these cases, connecting links are not found in the Amazonian basin. Some fish species show also disjunct vicariance with species in the small rivers of the Atlantic Rainforest.

Among the decapod crustaceans, the Palaemonidae prawns and the Trichodactylidae crabs, there are also several endemic species. The mollusc are represented mostly by widespread species of the La Plata basin, but there exists an endemic monotypic genus of the Pantanal, namely the hydrobiid snail *Aquidauania*.

In general it is a safe guess, that among the aquatic macrofauna, the percentage of endemics is markedly higher than among the terrestrial taxa. I would suggest a tentative rate of 10% of endemics. However, one has to take into account that these endemics are as a rule common to the whole basin of Rio Paraguay. The existence of a relatively large number of aquatic endemics, some of them of evidently ancestral taxa, has to be brought in accord with the hypothesis which assigns to the Pantanal only such a young, Pliocenic age.

The idea of a Mesozoic "Archiplata" (see above) is for instance used by Jaeckel (1967) in order to explain the high degree of specificity of the La Plata "endemics" among the mollusc. Noodt (1967) considers the existence of such a paleo-continent or archipelago, as a necessity to explain the many endemic subterranean crustaceans found in the extra-andine South America. For Jaeckel (op. cit.), Archiplata extended north, beyond the present Pantanal, to the basin of Rio Mamoré (=Guaporé).

The richest aquatic invertebrate fauna of the Pantanal is connected with the underwater roots of the floating and free floating plants. Although known only in part (Heckman, 1994; Rocha and Por, in press), the most obvious inhabitants of this environment, like the amphipod crustacean *Hyalella azteca* and the conchostracan crustacean *Cyclestheria hislopi* have an extremely broad distribution, that of the last species including also Africa. It seems evident that this fauna has the same invasive behaviour as the free floating plants among which it lives.

As mentioned above, the fauna of the Pantanal is incompletely known. Whereas the faunal lists of the vertebrates are fairly complete, more extensive knowledge exists only about the mollusc, the decapod crustaceans, the lepidopterans and to an even lesser degree the odonates (Brown, 1986). There exist already some data about the aquatic meiofauna too, but data on the terrestrial arthropods and other small invertebrates are too sketchy to be presented here.

21. Aquatic meiofauna

* *Zooplankton*

There is some information about the zooplankton of the Nhecolândia lakes, Lake Jacadigo and a few other areas. Rotiferans in two baías near Poconé were listed by Turner and Da Silva (1992). Among the 174 species found, most were cosmopolitic species. However, these authors consider a few species to be regional endemics: *Brachionus donneri*, *B. caudatus inusetus*, *B. dolabratus*, *Horaëlla thomassoni*, *Lecane(Monostyla) copeis*, *Platyias leloupi latiscapularis*, *Testudinella ahlstromi* and *Trichocerca chattoni*. The authors emphasize the predominant role of genus *Lecane* in the Neotropics; from this genus alone, 36 species have been reported by them from the Pantanal.

Kretzschmar et al. (1993) report seasonal aggregations of *Asplanchna sieboldi*, especially in the nearly drying-out pools of the late dry season. A whole list of other rotiferans characterize the nutrient rich waters of the early winter dry season. Kretzschmar et al. (1993) also mention that in the area of Poconé the flooded várzeas are nearly devoid of nutrients and of phyto- and zooplankton. The stagnant waters left behind by the floods are rich in rotiferan plankton. When the young fishes penetrate into the baías with the rising waters, these authors assume that predation of the rotiferans is very heavy.

A *Brachionus* sp. is reported as an occasional mass organism in the salinas of Nhecolândia.

The Cladocera of the Nhecolândia lakes, as reported by Mourão (1990) are fairly diversified. There are 5 species in the freshwater Baía do Jacaré (see above), namely *Daphnia* sp., *Diaphanosoma sp.*, *Ilyocryptus spinifer*, *Moina minuta*, and *Pseudosida sp.*. This composition rather resembles a phytal association. In the Baía do Arame, in which the floating aquatic vegetation was restricted, more planktonic species appear, such as *Bosmina longirostris*, *Bosminopsis* and *Ceriodaphnia cornuta*. An unidentified macrothricid cladoceran is reported from the salinas by Reid and Moreno (1990).

The copepod fauna in the Nhecolândia lakes is poor (Mourão, 1990). The cyclopoids *Thermocyclops minutus* and *Microcyclops* sp. are found in the freshwater lakes, whereas *Metacyclops mendocinus*, together with the above-mentioned *Brachionus* and a macrothricid, are the only meiofauna of the salinas.

A large collection of copepods has been studied by Reid and Moreno (1990). These authors report 19 species of copepods, from Rio Paraguay, Lake Jacadigo and a number of baías from Fazenda Nhumirim (Nhecolândia). The three species of harpacticoids reported, do not belong evidently to the plankton.

Four species of truly planktonic diaptomid calanoids are discussed by these authors. With two previous records, the diaptomid fauna of the Pantanal is considered poor as compared with the Paraná, the Amazon and the Orinoco basins.

The large and blue-coloured calanoid *Argyrodiaptomus* sp. appears only in the fishless baías (see below) and the authors believe that predation excludes them from other waterbodies. For this and the three species of *Notodiaptomus*, Reid and Moreno (1990) discuss the possibility of niche partition among the differently sized species and the populations of them. *Notodiaptomus coniferoides* is the sole calanoid present in Lake Jacadigo. It can be considered also to be a "true potamoplankter", frequently encountered in Rio Paraguay.

Concerning the cyclopoids in Reid and Moreno's study, it has to be emphasized that *Thermocyclops minutus* is the most numerous copepod in the area.

The largest of the three species of *Mesocyclops*, namely *M. longisetosus* has also been collected in two fishless baías. *Mesocyclops meridionalis* is a species which has been known until recently only from Corrientes in Argentina.

* *The phytal*

The rich radicular mass of the floating vegetation provides an ambient in which plant-connected aquatic fauna is extremely abundant. If, as suggested by limnologists, the primary production of the floating plants is in the Pantanal much more important that the primary production by the phytoplankton, without doubt the "secondary production" by the phytal fauna is as important.

Lately, this environment, the classical "hypopleuston", called by Heckman (1994) "interrhizon", is starting to attract the due attention. Heckman (op. cit.) emphasizes the role of the desmidian algal Aufwuchs of the roots and dwells on the rich insect fauna. He mentions the abundance of oligochetes of the genera *Dero* and *Aelosoma* and dwells chiefly on the aquatic coleopterans. According to this author, in his samples collected in the Poconé area, the coleopteran larvae *Suphisellus grammicus*, *Tropisternus collaris* and *Derallus rudis* predominate.

A recent research has been undertaken by C. F. Rocha and F. D. Por on the fauna of the floating *Eichhornia*, *Ludwigia* and *Salvinia*. in the southern Pantanal, and especially on Rio Miranda. A preliminary list of phytal cladocerans identified by H. Bromley-Schnur (Jerusalem), includes *Leydigiopsis ornata*, *L. ornata*, *L. curvirostris*, *Chydorus* cf. *ventricosus*, *Oxyurella longicaudis*, *Euryalona fasciculata*, *Indialona globosa*, *Duhnevendia odontoplax*, as well as *Ilyocryptus* cf.*sordidus*, *Diaphanosoma* sp., *Simocephalus* sp., *Alonella* sp., and Macrothricidae gen.

The lack of Bosminidae and of *Ceriodaphnia* distinguishes the phytal cladoceran assemblage from the zooplankton.

The copepod fauna of the phytal, as preliminarily identified by C.F. Rocha (São Paulo), contains *Macrocyclops albidus*, *Eucyclops neumani*, *Paracyclops fimbriatus*, s. l., *Ectocyclops phaleratus*, *E. rubescens*, *Mesocyclops longisetus*, *Microcyclops anceps*, *M. ceibaensis*, *M. varicans subequalis*, *M. finitimus*, and *Metacyclops*, sp.

The harpacticoid copepods occupied a much less important role, being represented by several species of *Attheyella* sensu latu. As a fundamental difference from the zooplanktonic assemblage mentioned above, there are only relatively few Diaptomidae in the phytal copepod samples.

Ostracoda compete with the Cladocera for the second place among the phytal crustaceans, the first place going to the Copepoda. There are probably up to 15–20 ostracod taxa, sometimes 5–6 in one sample. Unfortunately they are not identified yet. Ostracoda, though in relatively low numbers, were dominant in a sample taken from the centre of a camalote, where lack of oxygen prevails.

The conchostracan *Cyclestheria hislopi*, a faithful companion of the floating plants elsewhere, appeared in the summer samples analyzed from the Pantanal, only in relatively small amounts. Amphipoda made also a very spotty appearance.

The aquatic insects of the floating plants were very diversified. Ephemeroptera were always present in a low diversity, but with a tendency to be prevalent among roots of fixed plants. Odonata, both Zygoptera and Anisoptera, were present only in small numbers. Trichoptera were common in all the samples, represented by *Oxyethira* sp. and other Hydroptylidae. The larvae of Diptera were represented by Chironomidae and Ceratopogonidae, with the clear predominance of the first family. Chironomidae exhibited an apparent preference for the roots of *Salvinia*. The pelagic Chaoboridae appeared in the samples only accidentally. Hemiptera were exceedingly common, represented by the families of the Corixidae, Pleidae and Belostomatidae.

Here is the place to mention, that Corixidae live in enormous concentrations in the salinas of Nhecolândia, probably being the sole beneficiaries of the high densities of Cyanobacteria prevailing there.

Among the many Coleoptera larvae, the families Dytiscidae and Hydrophilidae stand out, although there were also a few Noteridae and Dryopidae. The larva of the dytiscid *Hydrovatus* is very typical for these environments.

The phytal fauna contains also Hydridae, turbellarians, bdelloid rotiferans, oligochetes glossiphonid hirudineans, oribatid acarians and hydracarians. Young stages of gastropod mollusc were frequent. Interestingly, the nematodes were almost absent in the summer.

It is worth mentioning that the phytal fauna is much richer and more diversified in the dry winter: nematodes, conchostracans, and harpacticoid copepods for example, are much more common than in the flooded summer conditions. Diaptomid copepods are also much more numerous than in the summer.

The meiofauna collected by Dioni (1967) in free floating plants on the middle Paraná, showed a somewhat different composition. The amphipod *Hyalella* was common. Among the dipteran larvae, Stratiomyidae were common. Hemipterans and conchostracans were not reported altogether.

* *Soft bottom meiobenthos*

Fukuhara and Mitamura (1985) calculated zoobenthos standing crops in Rio Miranda, Rio Paraguay and nearby baías and Fukuhara and Henry (1987) have calculated these values for an unnamed lake "near Cuiabá" (probably near Porto Jofre). The results are fairly inconclusive. While in both areas the chironomids are dominant and the ephemeropterans are surprisingly (for the authors) dominant, bivalves are lacking altogether in the first report, whereas in the second they appear with the highest biomass. Conversely, the oligochetes are dominant in the first set and absent from the Cuiabá lake. In general, the authors point out the very low standing crop of non-molluscan invertebrates on the bottoms.

22. Macrobenthos

* *The molluscs*

The Pantanal has possibly the greatest diversity of aquatic mollusc, especially of bivalves of South America. In this it is unparalleled by Amazonia, where acid blackwaters and extremely electrolyte-poor clear-waters are avoided by molluscs. There are many taxa which are endemic for the Paraguay-Paraná basin. The data on the mollusc are very scattered in old literature. I used the monographic paper by Jaeckel (1969) as well as the literature surveys by Paraense (1981), by Figueredo Alvarenga (1981) and the taxonomic revision of the freshwater bivalves by Parodiz and Bonetto (1963).

The dominant gastropods of the Pantanal are the Ampullariidae (Mesogastropoda), a typical Gondwanian family. The globose *Pomacea* (example *P. escalans*) and *Pomella* are characterized by the fact that they lay their calcareous eggs out of the water. *Asolene* (example *A. platae*), another globose form, lays gelatinous egg masses in the water. This is also the case with *Marisa* (example *M. planogyra*), a flattened "planorboid" genus. *Pomella* and *Pomacea* can move outside the water. *Pomacea* and *Marisa* can fill their mantle with air and become almost weightless in the water: they use this capacity in order to creep head-down under the surface film or on extremely thin plant fronds (Berthold, 1991).

Among the several species of Rissoidae (Hydrobiidae), there is a genus, *Aquidauania*, endemic to the Pantanal. There are also several species of the small hydrobiid genus *Littoridina*.

The Planorbidae (Pulmonata) are represented by *Drepanotrema* and by *Biomphalaria* which has been eventually introduced (by the slave trade) from Africa. *Biomphalaria* is a potential host for bilharzia worms. More adapted to the running waters is the hood-shaped ancylid *Gundlachia*.

As seen below, several species of birds are specialized feeders on the Pantanal gastropods.

The Lamellibranchia are almost all members of the South-American Hyriidae (Unionacea) and Mycetopodidae (Mutelaceae), Gondwanian families of the Unionaceae. The elongated *Diplodon* (example *D. paranense*) and the rather triangular *Castalia* belong to the subfamily Hyriinae. According to Heckman (1994), *Castalia ambigua* is the most common shell in the river banks and bottoms of the northern Pantanal. Several genera belong to the subfamily Mutelinae, like *Iheringiella* (example *I. balzani*), and *Fossula* (ex. *F. fosculifera*), with rather rounded shells; *Anodontides* (ex. *A. iheringi*) and

Leila (ex. *L. blainvilleana*), rather elongated and *Mycetopoda* (ex. *Mycetopoda legumen*) and *Lamproscapha* (ex. *L. ensiformis*) with very elongated and truncated shells. It is worth mentioning here, that the Hyriinae, have a "glochidium" larva which attaches itself to fishes, whereas the Mutelidae have a "lasidium" larva, very different in shape, and with a not really known behaviour. The hyriid *Schleschiella* has a non-parasitic glochidium (Parodiz and Bonetto, 1963). Genus *Byssanodonta* (=*Eupera*) *paranensis* is a small bivalve of the Sphaeriidae, which attaches itself with byssus threads to the roots of *Eichhornia*.

Thus, the different bivalves of the Pantanal have interesting adaptations to spread over wide areas, either carried by fish, or by the floating vegetation.

* *Decapod crustaceans*

Without going too much into details, the decapod fauna of the Pantanal is first of all characterized by a large variety of freshwater crabs of the family Trichodactylidae ("caranguejos"). They belong to the genera *Trichodactylus* (examples *T. parvus*, *T. borellianus*, *T. camerani* , etc.), *Sylviocarcinus* (ex. *S. pictus*), *Dilocarcinus* (*D. pagei*), *Zilchopsis* (ex. *Z. sattleri*) and *Poppiana* (ex. *P. argentinianus*).

The crabs are extremely important in the aquatic food webs of the Pantanal and serve as an important food object to the caimans. They are also extensively caught and sold as fish bait ("isca") to the many sport fishers.

The prawns (pitús") are much less in evidence, with *Atya paraguayensis*, *Macrobrachium amazonicum* and several species of *Palaeomonetes*, a marine-brackish water genus.

23. Ichthyofauna

Some zoogeographic aspects related to the fish fauna of the Pantanal have already been mentioned above. There are probably somewhat more than the 405 species calculated by Marins et al. (1981). Recent studies tend to split-off new species for the Pantanal populations. Such is for example the case of the classic sponge-eating *Leporinus friderici* described by Bloch in 1794 (Carvalho, 1986). It is thus possible that the ichthyofauna of the Pantanal will differ from the Amazonian by more than the 40% mentioned by Gery (1969).

The ichthyofauna of the Pantanal is much poorer than that of Amazonia with its over 1,300 species of fishes. The reasons can be of course historical ones, but in the present reality it is important that the Pantanal is a single large interconnected basin, whereas the Amazon is a collection of individualized tributary river basins. Furthermore each of those basins have different hydrography (ex. Madeira), hydrochemistry (ex. Rio Negro), or history (ex. Tocantins). The Amazon basin contains also mountain headwaters. The Pantanal rivers are all dominated by a similar hydrography without significant

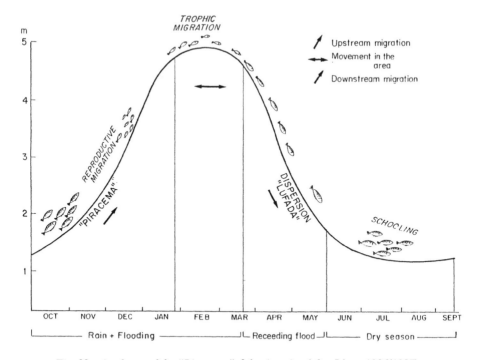

Fig. 39. A scheme of the "Piracema" fish migration (after Lima, 1986/1987).

separating rapids, they are all "whitewaters" in the Amazonian sense and all are lowland rivers.

There has not been an intensive co-evolution between "the fishes and the forest" (Goulding, 1980), like in Amazonia's blackwaters. Though there are many frugivores among the Pantanal characinids, the waters are highly productive and food is plentiful in the water. Also there are less and less dense flooded forests.

The "marine" element, though present in the Pantanal, is much less evident. The huge Amazonian *Arapaima* has been also probably unable to use the shallow headwaters contacts with the Pantanal (see above).

On the positive side, the fishes of the Pantanal present much more impressive migrations (see below) and the populations are probably larger. There seems to be a predominance of night-active fishes, which is probably induced by the very high predator pressure of the avifauna. The place of the giant *Arapaima* is taken by the local *Paulicea lutkeni* ("jaú"), which reaches a weight of over 80 kg. At the other extreme, there are the very small "lambaris". Among them *Hyphessobrycon flameus*, 2 cm long is probably the smallest Brazilian fish.

* *Piracema (Fig. 39)*

The seasonal migrations of the fish in the Pantanal collectively called "piracema", are a dominant and extremely impressive phenomenon, in which the most diverse taxa of small and large fishes participate. Hoehne (1936) was probably the first to describe the piracema which he witnessed during the Roosevelt expedition in 1913: "The fish seem to be taken by panic... they jump 1 to 2 m out of water... instantly we were hit by tens of fish and we had to protect ourselves with our arms in order not to be smothered...".

Indeed, being hit by 25–30 kg specimens, like a "pintado" (*Pseudoplatystoma*) or a "pacu" (*Piaractus*), must be a very traumatic experience.

The best analysis of the piracema in the Pantanal is given by Ferraz de Lima (1981; 1986/7) (Fig. 39). The adult fishes move up-river on the average, from May to October, i.e., during the dry period. In November-December with the raising waters, they reproduce and the fingerlings (together with the adults) penetrate into the flooded standing waters. This is a first critical phase, because often the flooded lowlands carry noxious products of decomposition of the plant detritus and are poor in oxygen. This is the phenomenon of fish death called "dequada". End of March and April, when the rains stop, the now well-fed young fish return to the rivers. This is the so-called "lufada", eagerly expected by many fish predators. During the winter the fishes assemble again in swarms and return up-river. The lufada is also a good period for lantern fishing ("pesca de facho") of several economically important species (see below). The massive outgoing migration of the giant *Paulicea* ("costada de jaú") is a very impressive sight.

The piracema migration takes the fishes over distances of hundreds of kilometres. There are no data from the Mato Grosso proper, but there are tagging results from Rio Mogi Guaçu and the Middle Paraná: curimatas (*Prochilodes*) can cover a yearly round trip of 1,200 km (Lowe-McConnell, 1987). Species identical with those of the Pantanal, travel yearly 400 km upstream on Rio Pilcomayo in Paraguay.

As typical for South America, the fishes belong mainly to the Characiformes the Perciform Cichlidae and the Siluriformes. There are, however, some very important less diverse fish taxa.

Among these one has to mention the freshwater stingrays *Paratrygon* with some Pantanal endemics; the "pirambóia" the very common lungfish *Lepidosiren paradoxus*, understandably well adapted to the drying out of the baías; the monotypic and endemic half-beak *Hypoclinemus*; the scienids *Pachyurus* and *Plagoscion*; the flatfish *Achirus* with endemic species and the needle-fish *Strongylura*, with endemics too. It is still a matter of discussion, how the freshwater fishes of marine origin reached the Pantanal: either as a result of a very old local evolution or through recent interchanges with the Amazon.

Another important "marginal" taxon is the eel-fish *Symbranchus vulgaris* ("mussum") representant of a monotypic genus widepread in South America. The mussum is a typical inhabitant and perhaps also an associate of the floating vegetation which vehiculates the fish over long distances.

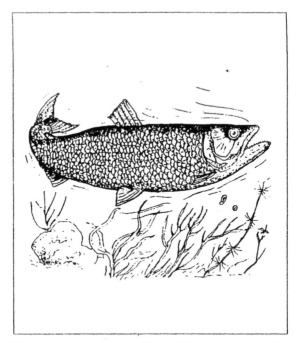

Fig. 40. The dourado, Salminus maxillosus (Vignette from Magalhães, 1992).

Electric eels of the Gymnotidae ("tuvira"), are also found in the Pantanal. Finally, the Cyprinodontidae, the tooth-carps, are represented by several species of Rivulinae, typically adapted to the life in extreme and temporary waters.

The characiforms are a very diversified lot of fish species and they can be only discussed in a very fragmentary way. Among the Erythrinae, the "traíra", *Hoplias malabaricus* is a most resistant fish, able to survive in the most adverse conditions in the salinas. It is a species widespread over the whole continent.

The relatively small *Prochilodes reticulatus* ("corumbatá") has been investigated for its migratory behaviour. It is assumed that the swarms of this fish cover distances of several hundreds of kilometres. Another typical piracema fish is *Brycon orbignianus* ("Piraputanga"). Its red coloured swarms are very visible in the rivers at migration time.

There are several species of *Parodon*, typical bottom feeders, living in the Paraguay system. A monotypic and very rare genus is *Roestes*, rediscovered not long ago by Menezes (1974) in Rio Jauru.

The "lambaris", various species of very small characids provided many very much appreciated aquarium fish, such as the "Mato Grosso", *Hyphessobrycon serpae*. There are many species of "tetras" and "neon tetras", as for instance the several species of *Moenkhausia* found in the Pantanal. *Cheirodon*

Fig. 41. The menacing teeth of a piranha, Pygocentrus nattereri (from Gery, 1969).

is not found in Amazonia. Another important aquarium fish is the silver hatchetfish *Gastropelecus sternicla* (Gery, 1977).

The bryconine characinid *Salminus maxillosus* ("dourado", Fig. 40) is a river fish with a remarkable trout-like shape and behaviour. It is very much appreciated by the fishermen. But perhaps the most sought after characinids are the discoidal and sometimes very large "pacus", species of *Collosoma*, *Mylosoma* and *Myloplus*. The most tasty pacu goes by its most recent name: *Piaractus mesopotamicus*.

Several characinids, like *Lebias* and the small *Pyrrhulina* live among the roots of the floating plants, taking advantage of the extra oxygen freed by the roots (Jedicke et al., 1989).

Sazima (1988) studied the agonistic behaviour of two characinids which hide too in the mass of underwater roots: the curimatid *Curimata spilura* feeds there on periphyton, whereas the serrasalmine *Catoprion mento*, a scale-eating predator, merely looks there for cover for its ambushing attacks.

The Serrasalminae, the "piranhas", typical predators, deserve special attention.

* *The Piranhas (Fig. 41)*

No South-American fish attained the notoriety of the savage piranhas. The Pantanal is much more infested by these fish than Amazonia. Some of them are stationary, non-migrant species, which do not leave the baías. The most common serrasalmids in the Pantanal are three species of *Pygocentrus*, three species of *Serrasalmus* and one of *Pygopristis* (Magalhães, 1992). The most dangerous is considered to be *Pygocentrus nattereri*, an eminently social predator. Other piranhas, like *Serrasalmus marginatus*, are solitary scale eaters. *Spilopleura* is also social, but less specialized as a predator (Sazima and Machado, 1990). Sazima and Zamprogno (1985) studied the close relation between young *S. spilopleura* and floating meadows of *Eichhornia* which serve both as hiding place and as feeding grounds. These authors consider that the "plague" of the piranhas is rafted over wide distances by the camalotes (see above).

The piranhas attack in swarms, which rapidly turn into hysterical shoals, if attracted by blood. Farm hands usually sacrifice an old cow, as a decoy for the piranhas, when crossing infested waters with their herds. A large mammal, like a capybara, can be skeletonized in a few minutes of frantic activity. Many a scientist working in the Pantanal suffered very deep bites from individual piranhas.

The caimans are efficient predators of piranhas and their demise because of the uncontrolled skin trade, usually results in a large increase in piranha populations. A dense cover by floating vegetation seems also favourable for piranhas. When caimans are present, they disturb the plant cover of the baías.

The catfish are fairly varied, but much less than those of Amazonia. The most colourful and also very sought after silurid is *Pseudoplatystoma corruscans*, "pintado", which carries different names according to the colour pattern. This is a fish of deep and calm river waters which often presents specimens with 25 kg of weight. The large *Paulicea lutkeni*, the 80 kg giants which can lay a single egg clutch of 6 kg weight (Magalhães, 1992) are also highly migratory riverine fishes. Small pimelodide catfish are also frequent, such as *Rhamdia pubescens* ("bagre"), *Pimelodus maculatus* ("bagre surubim"), small bottom living species of the backwaters. Among the loricariid catfish, there are several species of *'Plecostomus'* ("cascudo") which can cling to the rocky bottoms of the rapids in a typical loricariid way.

Finally the cichlids are also fairly varied, starting with the relatively large and economically important *Astronotus ocellata* ("cará açu") and *Aequidens paraguayensis* ("acará") and many tens of small species, such as the dozen or so species of genus *Apistogramma* and *Geophilus*. The cichlids are mainly associated with the aquatic vegetation and are preyed upon by other large fish, especially by the different piranhas.

Mourão et al. (1988) studied the diversity of fishes in the baías and salinas of Nhecolândia (see above). Their fish population depends of the environmental factors in the lakelets, but also of the frequency with which these reestablish connections with the rest of the Pantanal waters (Fig. 42). The most resistant fishes to salinity and/or insulation are *Hoplias malabaricus* and the small, plankton feeding *Pyrrhulina brevis australe*.

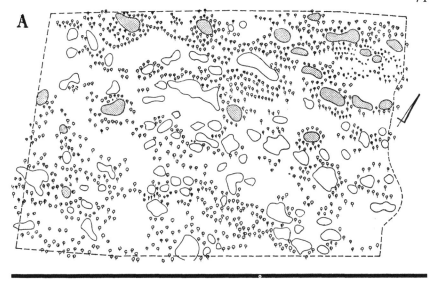
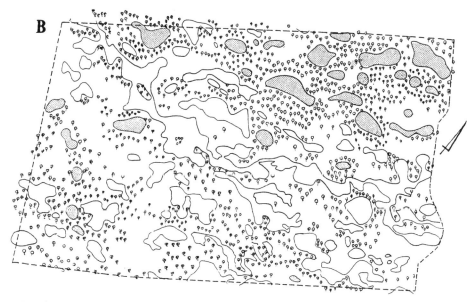

Fig. 42. Dry season and flood season situation of the baías and salinas of Fazenda Nhumirim (after Mourão et al., 1988). Grey-shaded waterbodies have reduced fish diversity or lack fishes altogether, because of hydrological isolation.

24. Pantanal fisheries

Fisheries are the second most important way of subsistence in the Pantanal, after cattle raising. The fishing devices are simple, with the predominance of angle-fishing in small boats. In the baías, the fishermen attract the fish by throwing corn into the water, the so-called "ceva". In turn, the herbivorous fishes attract the large carnivores.

The most exploited fish species "the noble fishes", are the curimbatá (*Prochilodes*), the pintado (*Pseudoplatystoma*), the dourado (*Salminus*), the jaú (*Paulicea*), the pacu (*Piaractus*), and more locally and occasionally species like the barbado (*Pinirampus*) and the piavuçu (*Leporinus*). In general these are rheophilous fish species. Recently, fisheries use freeze storage, and the export from the Pantanal increased geometrically (Pinto Paiva, 1984; Ferraz de Lima, 1986/7). Piranhas and other smaller fish are locally consumed. The piranha soup is a delicacy of the Pantanal cuisine.

The total marketed fish production of the state of Mato Grosso was (Pinto Paiva, op. cit.) of some 3,500 tons/year and of Mato Grosso do Sul of around 2,000 tons/year. But the figures are more than doubled, if illegal commerce is also considered. The official quantities have increased since, and especially the quotas fished by the amateur sport fishermen have increased considerably.

There is a general interdiction of fishing during the piracema season, on the average between November and January. However, the timing of the piracema is different in different rivers and different years. During this period only line fishing is allowed. Also there are fixed minimal sizes for the most important fishes.

The authorities tried during the last 20 years to impose individual quotas to sport fishermen. Today every person can take home 20 kg of mixed fish and one more specimen of unlimited weight. These rules are enforced by inspectors on the main access roads of the Pantanal. But smuggling is going on through many uncontrolled exits, including the countless small private aircraft.

With the development of sports fishing a new trade, that of the "isqueiros", the sellers of fish-bait sprung up. The isqueiros collect and sell large quantities of crabs and small fish and even of *Lepidosiren* lungfish. They are working mainly in closed baías in which the danger of being attacked by piranhas is great.

In the early 80's an Amazonian fish the tucunaré *(Colossoma macropomum)* was introduced into the headwaters of Rio Cuiabá. It is a valuable fish, but its expansion and interaction with the local fish fauna has to be monitored. after no less than five years, this fish already reached the Taquari system (Rezende, 1989).

25. The herpetofauna

The Pantanal is surprisingly poor both in diversity and in quantity of amphibians. It does not present the wealth of shaded wet environments, the many tanks of the epiphytic bromeliads, the small forest streams in which the diversity of the Neotropical rainforest amphibians is found. In the waters, the predatory pression by swamp-living serrasalmids and caimans and by the enormous flocks of waterfowl is probably responsible for that. The information on the amphibians presented below, is based mainly on the monograph by Cei (1980).

I did not find reports about caecilid amphibians in the Pantanal, although the presence of *Siphonops* in the Paraguayan Chaco has been reported.

The frog fauna is dominated by Chaco species with an admixture of northern wet-tropical species. The microhylid *Dermatonotus*, a burrowing termite feeder and the toad ("sapo") *Bufo paracnemis* (also called "cururu açu"), a very large animal of 20 cm length, and *Pseudis paradoxus*, with giant tadpoles, are Chaco species. To the same biogeographic province belong several leptodactylids, such as two species of *Physalaemus* ("rã de brejo"), which build large foam rafts on the water, on which reproduction and hatching occurs and species of *Leptodactylus* ("rã pimenta"). The species *L. labyrinthicus*, *L. fuscus*, and *L. mystacinus*, are mainly aquatic species. Among the small species *Phyllomedusa hypochondrialis*, is also a chacoan species. The pseudid frog *Lysapsus limellus* is probably the most aquatic among the Pantanal frogs, and is associated with the floating vegetation.

The wet tropical faunal element is less numerous and represented by the leptodactylid *Pseudopaludicola ameghini*, and by a few species of *Hyla* ("pererecas"). The Hylidae of the Pantanal are the most eurytopic and omnivorous representatives of the tree frogs, always typical for closed and wet forests. Some frogs of the Pantanal have poisonous or irritating cuticular excretions. Among them *Physalaemus biligonigerus* is producing a polypeptide, the physalaemin, which has an important medical use as a powerful hypotensive drug.

The reptiles of the Pantanal present also a relatively low diversity. As a rule the aquatic or amphibious species predominate. Cerrado species, mix with Paraná species and there are also some Amazonian influences.

One of the most remarkable aspect of the Pantanal is the abundance of caimans. Three species were recorded from the area. *Paleosuchus palpebrosus* ("jacaré coroa") is a northern species, present in Amazonia and also in northern South America. Its presence today in the Pantanal is doubtful, although it was still found at the beginning of the century in Rio Jauru' and Rio Paraguay by Perraca (1904). It has been probably extinct by overhunting. Another species, namely *Caiman latirostris* ("jacaré de papo amarelo"), a southeastern species seems to be also extinct in the area.

The third species, the paraguay caiman *Caiman yacare* ("jacaré" or "jacaré tinga"), probably the native Pantanal caiman, is still massively holding its ground.

Fig. 43. Caiman vacare (from Anonymous, album "Pantanal é Vida").

Fig. 44. Evening on a baía, with caimans (from Sucksdorff, 1987).

* The Jacaré (Figs. 43 and 44)

Caiman yacare is a relatively small species, reaching somewhat over 2 m, but usually between 80–125 cm. This is a southern species, known also from Bolivia, Paraguay and Argentina. In Paraguay it follows Rio Otoquis till near Santa Cruz. Another population lives in the swamps of Rio Mamoré and Rio Beni. The species has been reported also from Rio Guaporé (Medem, 1983).

Because of its important ecological role in the Pantanal and of the danger of poaching, the caiman is one of the main objects of management efforts in the area.

The caiman follows the Paraná southward and Krieg (1936) reported cases of jacarés, reaching the La Plata estuary on floating camalotes.

The jacaré passes long hours of the day outside the water, but feeds chiefly in the baías. Its hunting behaviour has been studied by Schaller and Crawshaw (1982) and Olmos and Sazima (1990).

Even if females can be aggressive in the brood-caring period, the jacarés present no danger to humans. Only bathers may be hit by "chicotadas" of caiman tails (Magalhães, 1992).

Their food consists mainly of gastropod mollusc and different fishes. Insects play an important role, especially in the young specimens and in the salinas (Coleoptera and Hemiptera, Santos et al., 1994). The trichodactylid crabs play also an important role. Besides, the jacaré is not choosy in its food, and other reptiles, birds, capybaras and other mammals are also found in their stomach contents. Cannibalism is also frequent.

Among the fishes found by Schaller and Crawshaw (1982) in the stomachs of the jacaré, *Plecostomus plecostomus*, *Hoplias malabaricus*, *Aequidens* sp., *Astronotus ocellata*, *Serrasalmus spilopleura*, *Brycon hilari*, and *Rhamdia* sp. are common, with a whole series of other smaller fish.

The literature is full of the role of the jacarés in controlling the populations of piranhas, or "palometas" as they are called in Spanish (Medem, 1983), but the fact have still to be substantiated. In some areas of Bolivia, there have been reports that the venomous stingray populations increased after the extinction of the jacarés. Medem (op. cit.) mentions also their role in controlling the populations of snails. Since these are primary hosts to flukes, cattle eating mollusc suffers from zoonoses.

Evidently, the baías from which the caimans were hunted, present changes more plant cover and the oxygen content of the water probably declines. The piranhas are known to be among the fishes which resist to low oxygen concentrations.

The jacaré populations live in loose colonies which are in principle dominated by elder males. The social structure becomes especially visible during the dry season, when many caimans aggregate in and around the few shrunken lakes. Interestingly, some lakes are over-populated, while others are empty of jacarés. Although they can survive in the salinas on an insect diet, it seems that the distribution is limited towards the Chaco by the increasing salinity of the waterbodies.

The jacaré in the Pantanal reproduces from late August to mid-April, i.e., chiefly during the wet season. The females build and attend the nests. Nest predation by coatis (*Nasua nasua*) and by the forest dogs (*Dusicyon thous*) is the most important factor in egg loss. Flooding of the banks and disturbance of nest-attending females by hunters are also important in lowering breeding success.

Medem (1983) reports rates of 77% and 54% of nest destruction and Crawshaw (1991) compares the breeding success in the more disturbed site near Poconé with that of a less disturbed one near Miranda.

Poaching on jacaré became a scourge in the later 1960's, when most of the other easily available species of extra-Amazonian crocodilians were already depredated to the limit of commercial expediency. In Paraguay, the species became near extinct, after yearly exports of caiman skins of up to 178 tons. In the then state of Mato Grosso, between 1965–1968, over 1 million skins were sold. After imposing controls, Bolivia still lost by smuggling into Paraguay some 300,000 jacaré skins every year. In 1980, the Brazilian Institute of Forestry (IBDF) still reported an extraction of about 30,000 jacaré skins per month (Dourojeanni, 1980). With strict control imposed in the Brasilian Pantanal, there is still occasional smuggling of skins to the Paraguayan town of Bahia Negra. During the decades of the 60th and 70th, the "coureiros", the skin poachers, ruled the Pantanal swamps, heavily armed and hand-in-hand with the drug smugglers. In the 80th, some 800,000 caimans were slaughtered yearly, from a total population estimated at 10 millions, and sold at 40 $ a piece to the skin traders. Today many "repented" coureiros turned into tourist guides. Presently CITES controls the trade in caiman skins and the Brazilian authorities closely cooperate. Recently several caiman farms have sprung up based first of all on collecting the about 70% naturally laid eggs, which otherwise would be lost to predation (Alho, 1986).

Another remarkable reptilian of the Pantanal is the caiman lizard *Dracaena paraguayensis* ("Vibora do Pantanal").

This semiaquatic lizard is a giant among the Teiidae, which can reach 1.5 m. The genus is monospecific and the species is endemic to the Pantanal area (Vanzolini and Valença, 1966). It is a specialized mollusc eater and its flattened strong crushing teeth and specialized maxillary joint are unique among the lizards. *Dracaena* is not venomous, as erroneously believed ("vibora"= viper!). These giant swimming lizards have also suffered from skin poaching.

Other lizards of the Pantanal, such as species of *Iguana* and the teiid *Ameiva ameiva* ("calango verde") (Magalhães, 1992) belong to the cerrado fauna. Another teiid *Tupinambis teguixim* is a very widespread lizard in central southeastern Brazil and in neighbouring countries.

Among the ophidians there are many interesting semi-aquatic species. Very few of the ophidiofauna are venomous (Magalhães, op. cit.). Two boide snakes are outstanding: the "sucuri", *Eunectes notaeus*, which can attain 7 m in length is an aquatic hunter, whereas the "jibóia", *Boa constrictor* is definitely terrestrial and found on higher grounds. The sucuri is still, frequent in the Pantanal, and represents one of the important tourist attraction of the region. A variety of colubrid snakes are aquatic and semi-aquatic ("cobras d'água"), such as *Liophis miliaris*, *Helicops angulata*, *H. leopardina*, *Hydrodynastes giga* and *Mastigodryas bifossatus*. The last two, called "boipevuçu" can bite very painfully, though they are not poisonous (Magalhães, 1992).

H. leopardina is a species typical for the Paraguay and the middle Paraná (Brown, 1986).

Finally though, the viperid *Bothrops neuwiedii*, variously called "jararaca", "boca de sapo" or "bocuda", is the dangerous venomous snake of the Pantanal.

The cerrado turtles *Geochelone* are common. Among the terapins *Platemys macrocephala* is possibly a Pantanal endemic (Brown, 1986) whereas *Phrynops geoffroanus* is an aquatic chelonid found also in southeastern Brazil.

26. The avifauna

Undoubtedly the most outstanding and lasting impression the Pantanal makes on a visitor is that of a paradise of waterfowl. Probably nowhere in the world can such a diverse bird life, be so easily observed and in such impressive populations.

It will be very difficult to give here a comprehensive picture of the birds of the Pantanal. Brown (1986) analyses the distribution and biogeographical affinities of over 650 bird species of the Pantanal and Dubs (1992) counts nearly 700 species. It seems evident that in our ecological treatment, the water- and wetland birds are of importance first of all. As a rule among the terrestrial birds we find species which are typical for the cerrados at large and of the tropical forests. Therefore from among them I shall mention only those which are especially in evidence in the Pantanal or are endemic to the area. The selection of some 100 species made by Magalhães (1992), an excellent and didactic observer will be of much help.

* Aquatic and wetland birds

The grebe *Podiceps dominicus* ("mergulhão pompom") is widespread in the Pantanal, with two other occasional species of grebes. The cormorant *Phalacrocorax olivaceus* ("biguá") is widespread and an extremely frequent typical colonial rooster. The "bigutinga" *Anhinga anhinga* is much less frequent. Among the 17 species of Ardeidae reported, *Ardea cocoi* ("garça cinza"), a southern element, and *Cosmedorius albus* ("garça alba") are important colony forming species. The capped heron *Ptherodius pileatus* ("garça real") is a solitary species.

Among the night herons, the boat-billed heron *Cochlearius cochlearius* ("arapapá") is a Neotropical colonial species, whereas the widespread *Nycticorax nycticorax* ("socó dorminhoco") and *Tigrisoma lineatum* ("socó boi") are solitary species.

There are three species of storks in the Pantanal. The most visible is *Jabiru micteria* ("tuiuiú"). It is the symbol bird of the Pantanal (Fig. 45). Tuiuiús

Fig. 45. Tuiuiú's, Jabiru mycteria (from Anonymous, album "Pantanal é Vida").

are very large storks, usually feeding in pairs in the shallow flooded areas. They also nest in pairs, with both parents taking alternatively charge of the chicks. The nest itself is a large structure of thick branches, used during several successive years. Dubs (1992) mentions the fact that these nests are usually inhabited also by the monk parakeet *Myopsitta monacha* ("caturnitá"), a aquatic plant-feeding species.

Mycteria americana ("cabeça seca") is a gregarious stork which has collective nests in the rookeries called "ninhais" (see below). Much more rare is the maguari stork *Ciconia maguari* ("joão grande") which builds large solitary nests on ground.

There are six species of ibises ("curicaca") as a rule colonial roosters. Among them *Phimosus infuscatus* is often found in large numbers. *Theristicus caerulescens* ("curicaca cinza" or "curicaca chumbo") is an endemic of the Paraguay-Paraná basin and a rather specialized mollusc feeder.

Finally, the roseate spoon bill *Ajaia ajaja* ("colhereiro") is fairly common, chiefly on the shores of the saline baías. This is a colonial bird which is facing extinction in the rest of Brazil and only the Pantanal maintains big populations (Magalhães, 1992).

Fig. 46. A "ninhal" nesting colony (from Anonymous, album "Pantanal é Vida").

Fig. 47. Air photograph of a group of ninhais (from Sucksdorff, 1987).

Fig. 48. A "dormitório", night roosting place.

* Ninhais

One of the most spectacular sights in the Pantanal is that of the rookeries on the communal nesting trees, the ninhais. These are heronries in which several nearby trees often shelter hundreds and thousands of aquatic birds (Figs. 46 and 47). Quite separate and different from this are the night rookeries, the "dormitórios" to which untold numbers of birds retreat at dawn (Fig. 48).

Ninhais in the Pantanal are being studied only recently (Ribeiro, 1987; Yamashita and Paulo Vale, 1990; Marques and Uetanabaro, 1993). As it happens with other aspects of Pantanal biology, research has been done on the ninhais of the Poconé and the Miranda regions. A ninhal usually represents a number of trees of the gallery forest, which are permanently or repeatedly occupied by the nesting species. The trees suffer extensively from the activity of the birds, more specifically they are burnt by the uric acid of the excrements, bleached and without leaves. Sometimes a tree becomes so fragile and so inadequate to give protection against strong winds, that the ninhal is abandoned and other unaffected trees are chosen. The position of the ninhais is defined besides the availability of trees, by the hydrography of the nearby baías and alagados during the high water season of the year. A ninhal can occupy a surface of several tens of hectares and contain several trees.

Several species are found together in the same ninhal, and the colonies usually number thousands of birds. The ninhal of Jofre Grande contained 12,000–15,000 birds in 1984.

Two types of ninhais are defined: 1. The white ninhais, more spectacular, in which the dominant nesters are *Mycteria americana*, *Cosmedorius albus* and *Ajaia ajaja*. 2. The black ninhais are basically composed by *Phalacrocorax olivaceus*, *Ardea cocoi*, *Anhinga anhinga* and *Nycticorax nycticorax*. The white ninhals are occupied in general during July–November and the black ninhais are occupied earlier and for a shorter period, i.e., between April and May. As soon as the youngsters are on wings, the populations disperse.

There is usually a certain spatial segregation between the different species on the same tree. For instance *Phalacrocorax* occupies the top, *Cosmedorius* the lower periphery and *Ajaia* are found in the centre and lower parts of the trees.

Naturally, these huge concentrations of birds attract a host of predators. Some catch eggs and chicks in the nests, such as howler monkeys *Alouatta caraya*, different raptorial birds and a variety of snakes. Other animals scavenge on the urine burnt ground, feeding on fallen eggs or chicken. Such are different quatis, *Nasua* and species of wolves. Also jacarés and piranhas partake in the feast.

Among the nine species of ducks on record, three are species of *Dendrocygna*. *D. bicolor* ("marreca-caneleira") used to be found in massive populations, but at present it became rare (Dubs, 1992). *Neochen jubata* ("pato corredor") and *Sarkidiornis melanotos* are Amazonian species found in the north.

Anas leucophrys ("marreca de coleira") on the other hand, is a typical duck species of the Chaco.

From the two species of screamers (Anhimidae), *Anhima cornuta* is a relatively rare northern species. On the other hand *Chauna torquata* ("tachã") is a typical Chaco element, the unique species of an endemic genus (Fig. 49). The first species feeds chiefly on insects and seeds whereas the second species eats chiefly aquatic plants. Usually posted on the top of the trees, the sharp alarm calls earned them the name "santinela do Pantanal".

From the many species of kites *Rosthramus sociabilis* ("caramujeiro") deserves special mention: this is a specialized mollusc-eating bird of prey. Its beak and claws are adapted to crushing and opening the valves and snails; around its nests often thousands of empty conches are found (Magalhães, 1992). Two other accipiterids namely *Busarellus nigricollis* ("gavião velho"), *Buteogallus urubutiga* ("gavião preto") are specialized aquatic predators and *Circus buffoni* feeds among others on frogs. The cosmopolitic osprey *Pandion haliaetus* ("aguia pesqueira") appears also in the Pantanal as an overwintering migrant (Dubs, 1992).

The limpkin *Aramus guatauna* ("carão") is also a specialized mollusc eater. Among the 11 species of rails ("saracuras"), aquatic habitats and aquatic food prevails. Most of them are also migratory birds. The cosmopolitic *Gallinula chloropus* ("frango d'água") and the beautiful *Porphyrula martinica* ("frango de água azul") often nest on floating plants. *Jacana jacana* ("jaçanã") is the classical bird adapted in every respect to living and nesting on large floating nympheaceans.

Fig. 49. Chauna torquata, the "Sentinela do Pantanal" (from Sucksdorff, 1987).

The strange sungrebe *Heliornis fulica* ("ipequi") is also a diving fish eater. This bird is famous for its intensive parental care (Dubs, 1992). The sun bittern *Eurypyga helias* ("pavãozinho do pará") is a fish and crustacean eating monotypic representative of a Neotropical family, famous for its colourful defensive display. Charadridae and Scolopacidae of the Pantanal are mostly migratory species. There are 5 species of kingfishers in the Pantanal (Brown, 1986). They are represented in general, by subspecies of widespread western hemisphere species (Dubs, 1992).

Of the two swallows, considered to be aquatic birds (Brown, 1986), only *Tachycineta albiventer* ("andorinha do rio") is considered as such by Dubs 1992) and by (Magalhães, 1992). Among the suboscine birds, three species of tyrants are also feeding on aquatic invertebrates and live near the water. The best known of them is *Fluvicola pica* ("lavandeira") (Magalhães, 1992).

* *Terrestrial birds*

The selection of species presented here will be even more restricted than that of the aquatic birds and we shall deal only with the most evident species. Already, like with terrestrial vegetation, the cerrado species predominate.

First of all the American "ostrich" *Rhea americana* ("ema") is very visible and widespread on permanent or temporary dryland. Almost as visible and impressive is *Cariama cristata* ("seriema") an nearly flightless tall bird, typical of the southern savannas of South America.

The heron *Bubulcus ibis* ("garça companheira") a worldwide associate with cattle, spread to the Pantanal too.

Extremely visible is the falconid *Polyborus plancus* ("caracará"). The caracaras live in pairs or solitary and feed on the open grounds, mainly of carrion but also of small live prey. In the modern Pantanal, they follow the open spaces created by the many dirt roads. They are important scavengers of the ninhals and of every roadside carrion. A small falconid, *Milvago chimachima* ("pinhé" or "carrapateiro") feeds on cattle ticks or even on occasional carrion (Magalhães, 1992). The urubu's *Cathartes*, with 3 species in the Pantanal, are usually less common than the caracará. The urubus live in large flocks and are more specialized carrion feeders. The ground living *Speotyto cunicularis* ("coruja do campo") is an important element of the cerrado fauna.

In the gallery forests one finds 8 species of cracids, some Amazonian and others southeastern in their distribution.

Typical for the cerrado are the two species of partridge-like *Nothura* ("codorna").

Parakeets are very common in the Pantanal and perhaps the most common and massive species is the "caturita" *Myopsitta monacha*, which feeds preferentially on Pontederiaceae. Among the macaws the blue *Anodorhynchus hyazinthinus* ("arara azul") is one of the largest and most beautiful parrots of Brazil. It is a cerrado species, very much endangered, especially since it seems to be feeding almost exclusively on the palm *Acrocomia sclerocarpa* ("bocaiuva"). Another endangered bird species of the Pantanal is the woodpecker *Campephilus robustus* ("pica pau rei") an endemic to the Paraguay-Paraná region. The largest species of tucans of Brazil, *Rhamphastos toco* ("tucanuçu") (Magalhães, 1992) is a nest-robber (Dubs, 1992).

Among the 29 species of hummingbirds ("beija flores") listed in the Pantanal (Brown, 1986), several species of *Phaethornis* are endemic to the Paraguay-Paraná region. Some of the hummingbirds are migratory.

A cuckoo, *Crotophaga ani* ("anu preto"), a black-coloured colonial nesting species, appears associated with the Pantanal cattle. Magalhães (1992) discards the view that they feed of cattle ticks and rather thinks that they perch on the large animals.

Among the many passeriform birds a few are swamp dwellers like tyrannids *Fluvicola pica* ("lavadeira") and *F. leucocephala* ("maria velhinha") (see above). The fringillid genus *Paroaria* ("cardeal") with three species, forms very numerous bands everywhere in the Pantanal, even in the vicinity of human dwellings. The beautiful red-headed birds have also been chosen as a symbol for the Pantanal (Fig. 50).

Fig. 50. Cardeal, Paroaria capitata (vignette from Magalhães, 1992).

The Pantanal is an important habitat for the Nearctic migrants. Brown (1986) lists 32 migrant species. Among the migrants, the Charadridae, Scolopacidae and Hirundinidae are predominant. But there is also the interesting case of the bobolink, the icterid *Dolichonyx oryzivorus*, which migrates to breed in North America, although it belongs to a Neotropic family. According to Sick (1983), three migratory routes converge on the Pantanal.

Several typical cerrado scavengers, like *Cariama*, *Polyborus* and several Tyrannidae, are "fire followers", i.e., they migrate according to the seasonally occurring fires (Barthold, 1993). See also the above-mentioned "gavião fumaça".

27. The mammals

According to Schaller (1981), there are some 64 species of mammals in the area of Serra do Amolar, on the big ranch of Acurizal, which he studied, a third of which are bats. Brown (1986) considers this to be a very representative sample of the Pantanal fauna. The great majority are widely distributed South-American open land species, though some are represented by local subspecies (Brown, op. cit.).

Furthermore, Brown quotes Prof. Cleber Alho's thumb rule according to which the small mammals are mostly Amazonian and cerrado species, whereas the medium-sized ones are Chaco species.

There are few outright aquatic mammals in the Pantanal. The large manatee *Trichechus* and the river dolphins *Inia* did not reach the Pantanal. The Mustellidae are represented by the ariranha *Pteronura brasiliensis* (also known in Amazonia) and the lontra *Lutra platensis*. The ariranha is a very large otter (up to 1.2 m long) and its glossy pelt is very much appreciated by the poachers. Yet it seems that at present, with stringent protective laws, the Pantanal populations of this species are not at risk any more. According to Magalhães (1992), the ariranha is preying on piranhas.

The most visible wild mammal of the Pantanal is the omnipresent capybara *Hydrochoerus hydrochaeris*, "capivara" or by its Spanish name "chiguire".

* *The Capybara (Fig. 51)*

This is the largest rodent of the world, often reaching a weight of 50 kg. Alho (1986) reports cases with over 70 kg. The capybara is a peaceful herbivore which can be considered to be semiaquatic. Owing to its webbed hindlegs, the animals are excellent swimmers and divers. Females can be often seen swimming at the top of an angle formed by the youngsters. When in danger, the capybaras take to the water. They usually even copulate in the water. Interestingly, the capybaras are collective breeders, with females often taking care of a whole creche of alien brood (MacDonald, 1981).

Capybaras live in social groups which contain in medium 8–16 specimens. In the hight of the calving season, they may count up to 49 specimens. The groups are family groups of one dominant and often a single male (MacDonald, 1982). They reproduce twice a year with a not well defined peak of prolificity in July-August. The densities of the capybara populations are impressive: during the flood season they may reach nearly 15 specimens per square kilometre of dryland and a biomass of 443 kg/square km (Alho, 1986).

According to Alho (op.cit.), the ideal environment for these animals contains a baía or corixo, with a cordilheira and a small fragment of gallery forest. The capybara is perhaps the only major macroherbivore of the Pantanal, since the demise to the large herds of deer and of course a concurrent of the cattle herds. A reduction in the number of the jaguars owing to hunting, contributed also to the increase in the capybara populations.

Much attention is being given to the parasitic diseases of the capybara. They suffer from heavy parasitic diseases. *Trypanosoma evansi* is the most important killer of these animals (the illness called "mal de cadeiras" (Magalhães, 1992)). They seem to transmit trypanosomiasis also to the different species of deer. The cattle breeders see in the capybaras also a carrier of *Brucella abortum*, the agent of the dangerous brucellosis. As a result of studying the parasitofauna of *Hydrochoerus*, Costa and Catto (1994) found in one population in Nhecolândia, no less than 12 parasitic helminth species.

Fig. 51. Capybaras, Hydrochoerus hydrochaeris (from MacDonald, 1981).

Capybaras are being hunted for their coat and for their meat. They are eaten at lent by devote catholics, because, being aquatic they are technically considered to be fishes. There exist today various pilot plants for capybara raising. According to Alho (1986), this species does not present any serious problems to captive cultivation, besides the trypanosomial disease.

However, it is still a fairly common practice by the ranchers to kill hundreds of capybaras, for supposed sanitary reasons.

The capybaras are extremely widespread in South America. Two other rodents, though much smaller than the capybara, are widespread too and also sought after by the hunters, because of their tasty meat. These are the "cutia" *Dasyprocta variegata* and the "paca" *Cuniculus paca*, a good swimmer. Among the many smaller species of rodents, *Nectomys* ("rato de água") is an aquatic species (Magalhães, 1992).

Among the ungulates, the diversity of the deer stands out. There are five species of cervids in the Pantanal, the largest concentration of Artiodactyla in South America.

Among them *Blastocerus dichotomus* ("cervo pantaneiro") is a true wetland deer, with legs adapted to thread on soft bottoms. It is the largest South-American deer, reaching a weight of 120 kg. It feeds on swampy vegetation, often neck-deep in water. The swamp deer is not jumpy as the other species and the males are not fighting among themselves (Magalhães, 1992).

Ozotocerus bezoarticus ("veado campeiro") is a deer of the open vegetation, a mainly dusk and night feeder, which has been very much overhunted in Brazil and even its populations in the Pantanal are endangered. There are three species of *Mazama*, all with unbranched antlers and all of them good swimmers. *M. americana* ("veado mateiro") is a forest species. The other two, *M. simplicicornis* and *M. guazoubira*, are small species, much less than 1 m big.

The deer of the Pantanal formed an integral part of the local culture of the Indians (see below). Today their stands are much reduced by hunting and by competition for food from feral cattle and of course from the herds of them. Arruda Mauro et al. (1994), calculated the existing populations of *Blastocerus* in the Pantanal in 1991 and reached a figure of 36,314 ± 4,923 specimens, which is not lower than the population calculated in 1974 by Schaller and Vasconcelos (1978).

There are two species of wild pig. *Tayassu tajacu* ("caitetu") is a small species of the forest edges and river banks; the caitetus are good swimmers. *Tayassu albirostris* ("queixada") is a social forest dweller, with very aggressive boars. A feral pig "porco monteiro" is also widespread. Alho (1986) considers that the wild pigs of the Pantanal are resisting well to hunting pressure. Finally the tapirs *Tapirus terrestris* ("anta"), typical forest dwellers, are also encountered in the Pantanal.

With such a rich prey in the Pantanal, it is not surprising that the diversity of the carnivores is large. Of the four wild dogs *Speothos vinaticus* ("cachorro vinagre") is the most typical, though very much endangered. It is a semiaquatic gregarious species feeding most of all of crustaceans and of the fall-off of the ninhais (see above). Another rarity is the wild dog *Chrysocyon brachyurus* ("lobo guará"), a typical cerrado species, endangered over all its range. The two other wild dogs are species of the genus *Dusicyon*. The jaguar *Panthera onca* ("onça pintada"), the large and unjustly feared predator, used to be very frequent in the Pantanal, feeding on capybaras, pacas and other medium-sized prey. It appears in different colour variations, a dark-pelted form called "panthera" and a smaller and small-spotted variety called "cangussu" (Magalhães, 1992). The "pantaneiros", the local inhabitants specialized in hunting the jaguar, traditionally with a single spear, more in order to protect their cattle than for self-defence. Although the over 100 kg cat can be dangerous, it never attacks men. Today, the jaguars are more and more restricted by the expanding cattle ranching, although not yet in danger of extinction. This danger is much more concrete for a series of small forest cats, the jaguarundi *Felis yagouaroundi* ("gato mourisco"), a thin and elongate climber, the margay *F. wiedii* ("gato maracajá")and the tiger cat *F. tigrina* ("gato de mato pequeno"). The ocelot, *F. pardalis* ("jaguatirica") and the puma *F. concolor* ("onça parda") are much hunted for the supposed damage they do(Magalhães, 1992).

There are two racoons in the Pantanal: the typical aquatic feeder *Procyon cancrivorus* ("guaximim") and the more terrestrial and easily domesticated coati *Nasua nasua* ("quati"). There are also several species of terrestrial Mustellidae.

The monkey fauna of the Pantanal is very impoverished. In fact only two monkeys are common. The howler *Alouatta caraya* ("bugio preto") a species of the cerrado woods, is very common in the gallery forests of the Pantanal. Equally common, but less visible is the small brown capucin *Cebus apella*, an extremely widespread South-American species.

There are several species of opossums in the Pantanal, belonging to genus *Didelphis* ("gambás") and "cuicas", belonging to other didelphid genera (Magalhães, 1992).

Typical cerrado species are the 4 species of pangolins ("tatus") and the two species of ant eaters. The great ant eater *Myrmecophaga tridactyla* ("tamanduá bandeira") used to be very common on the drylands of the Pantanal, but has been lately subject to extensive hunting. As many Pantanal animals, the ant eater swims very adroitly too.

Finally, mention has to be made of the bats, some 20 species of them. Most species are Amazonian ones which reach their limits in the northern Pantanal (Brown, 1986). Understandably the fishing bat *Noctilio leporinus* ("morcego pescador") is very common in the Pantanal.

As new additions to the mammalian fauna of the Pantanal one has to mention the feral pigs, feral cows, the peculiar Pantanal horses and the buffalos (see below). The feral animals, the so-called "baguá", are free-for-all hunting objects, and are also considered to be carriers of diseases.

The feral pigs became important meat suppliers to the pantaneiros. The practice is a typical managed hunting in which known populations in the wild are carefully culled.

28. The humans in the Pantanal

* The Indians

The Pantanal Indians are the epitome of the Brazilian aborigines. Unlike the coastal Tupis which were soon overwhelmed and driven to near extinction by the colonizing Portuguese and unlike the Amazonian Indians, protected and unknown in their forests till recently, the Indians of the Pantanal played an important historical role as late as the second half of the XIXth century. According to Padua Bertelli (1988), there were some 70 nations of Indians in the Pantanal. One of the circumstances which delayed the colonization of the Pantanal has been the valiant fight put up by the local Indian nations, in a way reminiscent (sadly!) of the Indians of North America.

The ruling Indians of the Pantanal were tribes and nations belonging to the Tupi-Guarani language group. These tribes are collectively called Mbayá. Other tribes, subject to the Mbayá belonged to the "Arawakian" farming groups, collectively called Guaná. Some tribes had other affiliations, like the Bororós of the eastern approaches of the Pantanal, and the very primitive Nambikwaras of the north.

The Indians developed a remarkable mastery of their environment, cultivating their staple crop, the wild water rice *Oryza perennis* (Zerries, 1968) and building artificial islands within the swamps (Hoehne, 1936). These islands can still be identified. Some of the tribes, like the Paiaguás and the Guatós were expert boatmen and their up to 6 m long dugouts served not only for fishing and transport of merchandise, but also for river piracy. The southern Guaicurus, after obtaining horses from the Spaniards, turned into fearsome horsemen and warriors.

The Indians developed a culture largely based on deer hunting (Padua Bertelli, 1988). Sometimes they hunted with fire corrals and sometimes also as single spear hunters, like the "zagueiros" of the modern pantaneiros.

The Pantanal Indians were advanced socially and culturally. The Mbaya's had a caste society with the nobles and priests at the top, then the warriors, the subdued field worker Guanás and finally the slaves captured in war.

For many decades, the Pantanal indians successfully repelled the European colonists. There are reports of many inter-tribal wars, like those of the Rio Paraguay tribes of the Paiaguás with the more northern Guatôs. During the first half of the XVIIIth century, the Paiaguás and the Bororós around Cuiabá killed and enslaved many Portuguese colonists. With the discovery of gold near Cuiabá', many convoys were sacked and the gold sent and sold downstream to the spaniards of Asuncion.

The Portuguese built during the later XVIIIth century a series of fortresses along the Paraguay and by the end of that century, the Bororós and the Paiaguás were virtually annihilated in the Pantanal. The Bororós though, still hold out on the Planalto (see Gomes de Souza, 1986).

Fig. 52. A Cadiveu woman with traditional body painting (from Levi Strauss, 1994).

The Guaicurús fought on and during the Paraguayan war (1864–1870), sided with the Brazilians against the Paraguayans, and earned official thanks and status of allies from the then Emperor of Brazil (Valverde, 1972).

Tribes of the Guaicurú nation still survive in reserves, the most important ones being the Cadiveus along the Serra da Bodoquena.

* Cadiveus

Claude Levi-Strauss (1976) visited this tribe in the 1930's and his description of their ethnography is today a classical of the so-called "structural anthropology". Still at that date, it was possible to study the social structure and the mores of a typical Mbayá Guaicuru tribe, although acculturated in many aspects. They had a very elaborate social and caste system. The Cadiveus were the master tribe, itself divided into various casts of nobles and of warriors. Below them, lived the Guanás, farmers of the Tereno tribe. The lowest class were the Chamacoco slaves, but also an admixture of escaped negroes. Levi-Strauss reported on the peculiar habit of controlled infanticide among the higher classes. On the other hand they practised large-scale child abduction and adoption from other tribes.

The very complicated exogamy rules of the various Mbayá casts found their expression in the beautiful body paintings on the face and even on the whole body of the woman. This art of the ritual and heraldic body painting reached its zenith in South America with the Cadiveus (Zerries, 1968). The hieroglyphic paintings symbolized genealogy and status (Fig. 52). The nobles of the Cadiveus had a status comparable to that of the chivalry stories. In Levi-Strauss's romantic words: "In this charming civilization, the female beauties

Fig. 53. Cadiveu pottery.

trace the outlines of the collective dream with their make-up... whose mysteries they disclose as they reveal their nudity".

Body painting was an exclusive trade of the woman, whereas the men produced sculptures. The Guanás added to this civilization, their exquisite trichrome pottery of the Marajoara style, which they brought with them from their northern Arawak homeland. During the 60 years which passed since Levi-Strauss' visit, the Cadiveus survived, probably loosing many of the ancient traditions and trades. However, their trichrome pottery survived into the tourist age (Fig. 53).

The Indians of the Pantanal found their way in the often romantic stories of the XVIII–XIX century travellers, and they were bucolically painted by the craftsmen of the various expeditions. The friendly and peaceful Nambiquaras on the Upper Paraguay were met and described by Theodore Roosevelt at the beginning of this century. More than half a century after his "Tristes Tropiques", Levi-Strauss published his old photographs of Indians in "Saudades do Brasil" (1994). It is to Levi-Strauss and his Pantanal encounters that we owe our present image of the peaceful, children-loving Brazilian aborigines, bathing and frolicking in the waters of the Pantanal. Even such a blasé modern novelist as John Updike, turns romantic in his recent "Brazil" (1994) when his heroes travel towards the heart of the Pantanal and at the same time back in time: they meet there, dreamlike, ferocious Guaicurus and the gentle Tupi-Kawahibs of Levi-Strauss.

On a more concrete side, Cândido Rondon, the soldier-politician-ethnographer, born in 1865 in the Pantanal settlement of Mimoso, was an arduous protector of the Indians. During his famous "Rondon Mission", in the first decade of this century, he made contact with many unknown Indian tribes on the periphery of the Pantanal, investigating them and bringing them, what he thought, the luminaries of civilization. In much more realistic times, we realize that this has been in fact the start of a process of degeneration and disappearance; but the later marshal Rondon, the first Brazilian "indigenista", was animated by the best ideals of those days.

* Colonization

As mentioned above, the Pantanal has been avoided by the Europeans for the better part of three centuries. Only in the XVIIIth century did sporadic forays of the "bandeirantes" armed slave-hunters of São Paulo, penetrated the Pantanal. Their first victims were the Bororós of the Cuiabá basin.

Everything changed when gold was discovered by the bandeirantes near the town of Cuiabá of today. The first protected overland road to São Paulo was opened in 1730. In fact, it was the need to protect this gold from the marauding Indians and ultimately from the Spanish enemy, that the first forts and later towns sprung up. Cuiabá was founded already in 1719, but had a difficult start. It was followed by the border forts of Corumbá in 1775, Cáceres in 1778, Poconé in 1783 and Miranda in 1793.

Next step in the colonization has been the carving-up of the Pantanal into various "sesmarias", donations of land made by the Portuguese crown. Such sesmarias measured in average around 13,000 hectares each (Valverde, 1972) and were often defined by river courses. The principal labour force used, were slaves, first Indians and than to a much lesser degree negroes. With the end of slave trade in the XIXth century, the sesmarias decayed and even those which avoided splitting up to the descendants through intermarriage, became essentially poor lands. When land acquisition through purchase turned legal, the way for the creation of the huge Pantanal estates became free. Landowners with properties of hundreds of thousands of hectares abounded. Tomas Laranjeira owned 5 million hectares. Fazenda Firme, between the Taquari and Rio Negro is today the subregion known as "Nhecolândia" after its most famous owner Joaquim Eugenio Gomes da Silva, nicknamed "Nheco". Even in this century, the "Fazenda Francesa" owned more than 600,000 hectares and today Fazenda Bodoquena measures 130,000 hectares. The settling of the Pantanal was interrupted by the Paraguayan war (1864–1870). Many Pantanal towns, such as Corumba and Cáceres were abandoned to the spanish invaders.

Although its last phases of the war were fought out in the southern Pantanal, near Miranda, the main effort went to the control of the Rio Paraguay shipping. It was there that important naval battles were fought. But after the steamship reigned the Paraguay for more than half a century after their first appearance there in 1857, the difficulties of navigating the complicated bends of the river, the "estirões" and the shallows, made this waterway less and less profitable. In 1913 the railway reached Corumbá on the Rio Paraguay, crossing into Bolivia. This was the first and only attempt of a transcontinental railway in South America.

Fig. 54. A herd crossing a river (from Anonymous, album "Pantanal é Vida").

During the troubled years of the war, much of the livestock was left in place by the escaping landowners. Much of the pig, horses and even steers turned wild. As an example the feral pigs survive till now. Herds of the Pantanal horse, a variety adapted to the swamps and bred by the Guaicurus turned feral and it is owing to these herds that the variety still survives in modern breeds. Like the famous horses of the Camargue in France, the Pantanal horse is a swamp-adapted animal.

With a very short interval in which Pantanal lands produced "ierba maté" for export, during the period of Paraguayan occupation and influence, the principal product of the Pantanal has been and still is the cattle raising (Fig. 54).

* *Nelore*

There is a saying that the cattle brought the white man to the Pantanal. In fact, the first herds were brought still by the Spaniards of the La Plata. A local variety evolved, named "Tucura" which is well adapted to the local conditions and resistant to pests (Dourojeanni, 1980). However, with the opening of the overland trail, in the XIXth century, the zebu cows appeared in the Pantanal.

Fig. 55. Drying of "carne seca" beef, in the 30's (from Levi-Strauss, 1994).

Today, there are chiefly two varieties of zebu, the "Nelore", the most widely raised and the "Gir". Brazil developed many local races of this Asiatic cattle.

The old Tucura race is still present and sometimes "baguá", feral cows are captured too.

The earliest way to commercialize the beef was by drying and salting (Fig. 55) and by producing meat extract. During this period which culminated in the First World War, Rio Paraguay was the main artery of commerce. After the completion of the rail road, and the advent of the modern refrigeration techniques, the herds were increasingly taken to slaughterhouses in the east. At times, the Pantanal held over 4 million heads of cattle. Single fazendas often own 30,000–40,000 heads of cattle. In the decade of 1970, between 100,000 to 200,000 heads left yearly for the slaughterhouses (Gomes de Souza, 1986). After a cycle of dry years, heavy floods which started in 1974 and 1977 and continued till 1982, severely curtailed the herds. Much pasture land was lost and hundreds of thousands of cows died or drowned. In 1974 alone, some 800,000 heads of cattle were lost. In the 1980's the export of cows was more than halved, to 30,000–60,000 heads yearly. Some solace was the introduction of cattle boats, "boieiros" which could load and save whole herds.

While the Pantanal was believed to held some 6 million heads before the last cycle of floods, it is now considered that 4 million would correspond more to the overall capacity of the region.

The Pantanal cattle is raised free. It spends all the time in the field, on natural food. Therefore only the huge pastures available, compensate for the very low productivity of the herds. This is the so-called lean cattle, "boi magro" of the Pantanal. On the other hand one can safely say that the Nelore cattle is integrated into the ecosystem of the Pantanal. Alho et al. (1988), however, present the view that an emphasis on the old Tucura cows, adapted to the

Pantanal for more than 200 years would be ecologically more sensible. The plans to introduce buffaloes on a large scale, is considered to be dangerous, because the trampling of the soil by these heavy animals.

With mammalian grazers reduced to the highly culled populations of deer, the possibility existed to accommodate the millions of ruminant newcomers. Today, the widely spread cows feed side-by-side with deer, ema and capybara. For those used to the African savannas, the Nelore occupies almost alone, the niche of the many species of hoofed animals which are grazing there.

* *The pantaneiros*

There are very few studies dealing with the modern population of the Pantanal. Many of the data exposed here are taken from Silva and Fernandes Silva (1992), on the model of the socio-ethnography of an area in the wider Cuiabá region.

Like everywhere in Brazil, the three basic components: Indians, Europeans and Negroes are represented. The difference is the presence of an important Paraguayan component, which finds its expression in the local culture. The Indians contributed relatively much, since their nations were still numerous in recent times. Perhaps the influence of the Bororós is among the most important. Indian slaves have been more important in the Pantanal economy than the negro slaves. The Europeans, are either Spaniards from Paraguay and Bolivia or Portuguese, coming until recently chiefly from the states of São Paulo and of Minas Gerais, The new European or Japanese immigrants which reached Brazil in this century did not have any marked influence.

There exists a basic differentiation into "ribeirinhos", which are fishermen or part-time fishermen and "pantaneiros", which are farmers. The fishermen, organized in colonies around communitarian fishing titles, are first of all descendants of Indians and of liberated Indian or negro slaves. The ribeirinhos live in "regular" established settlements along the rivers. Besides this, there is nothing in the Pantanal which can be compared to a village.

The pantaneiros had a different history. Crop growing has been till very recently, nothing more than a subsistence agriculture; the cattle became right from the beginning, their principal source of income. In the process of expansion of the cattle raising, the old "sesmarias", fragmented in small holdings decayed and the young generation moved to the towns. The large fazendas which sprang up, are owned generally by absentee landlords (see above). In 1980 there were already more than 1,500 private landing strips in the Pantanal. Most of the lands were left either as natural pasture, or planted with exotic African grasses, often even by air-seeding. The ranch hend, the "boiadeiros", became the dominant ethnographic group of the non-urban population of the Pantanal. The boiadeiro, with many trappings of a cowboy, with their lassoes and long bugles made of cattle horns (the "berrante"), are permanently on the move, following the cattle and driving it to the slaughterhouses. The boiadeiros are as a rule only a male population, some of them with families

Fig. 56. Boiadeiros, ranch hend with the typical bugle.

in nearby towns or rather far in eastern Brazil. During the zenith of the huge fazendas of the Pantanal, armed boiadeiros served as small private armies to defend the enormous ranches (Fig. 56).

The "coureiros" were active in the trade of animal skins, especially of the caimans. Even in the very last years, they were in a violent conflict with the authorities. In the early 1980's it even happened that young naturalists were killed by these poachers. Today the caiman skin "trade" nearly disappeared. According to many data presented by Alho et al. (1988) skin poaching, especially of the large cat species is still a danger.

Finally the "garimpeiros" are primitive gold prospectors, mainly washing the gold from the river sediments (see below).

Their number increased very much following the economical crisis of the last decade in Brazil. Unemployment sent also thousands of people into uncontrolled and predatory fishing to supply the many freezer boats and to service the hoards of individual "sports fishers". Like the garimpeiros, these unorganized fishermen are also a semi-nomad males-only population.

The towns of the Pantanal had a very slow and painful development. Many of them were evacuated and resettled after the short Paraguayan invasion. Only Cuiabá had a more harmonious, in Brazilian terms, history. It is a city which is approaching the 500,000 population mark. Cuiabá has all the trappings of a state capital (it is the capital of Mato Grosso state), and houses an active university. Corumbá the "white City", which is called "the capital of the Pantanal", has barely 100,000 habitants, is suffering still from an island-like isolation from the rest of the country and presenting still the features of a "Far West" frontier town. Its port, once bustling in the highdays of the steam boats of the Paraguay river, is now a historical relic. It is, however, worth

Fig. 57. Pantanal tracks in the 1930's. For many parts of the Pantanal there has been no qualitative technical improvement (from Levi-Strauss, 1994).

mentioning that Corumbá is the centre of a fledgling "autonomy" movement which speaks of a "Republica do Pantanal". It is linked by a precarious asphalt road to the rest of Brazil and the narrow-gauge railway which crosses from Corumbá into Bolivia is operated only twice weekly. Often the roads are not much better than those found by Levi-Strauss in the 1930's (Fig. 57).

The air service, usually excellent all over Brazil, is rather rhapsodical when it comes to Corumbá. Still, the recent waves of tourism brought in 1992 nearly 30,000 foreign tourists to Corumbá turning this sleepy town into the seventh most important foreign gate of Brazil.

A minor mining industry exists in Corumbá, using the iron and manganese ore of the Serra de Urucum. With the recent agreement for supplying Bolivian natural gas to Brazil, this mining industry is bound to develop and it is feared that industrial pollution will spoil many of the scenic sites.

The capital of the state of Mato Grosso do Sul, Campo Grande with more than 500,000 inhabitants is (happily) situated already beyond the watershed of the Pantanal.

The best example of the "new frontier" policy of the Vargas regime of the second World War times, has been in the Pantanal the planning and building of the "Transpantaneira " road, a distant and smaller-scale replica of the "Transamazonica" highway of unhappy memory (Fig. 58).

Fig. 58. A stretch of the Transpantaneira road (from Anonymous, album "Pantanal é Vida").

In this case, the end came to the project because of the splitting of the old state of Mato Grosso in two, in 1979. As a result, the Transpantaneira, starting from Cuiabá, stopped short before the border of the new state of Mato Grosso do Sul at Porto Jofre, and did not continue to Corumbá as planned. As every road which ends in a dead-end, this was the death warrant for this project.

29. Environmental worries

The size of the Pantanal itself wouchsafes for a certain protection, or "homeostasis" in the face of the dangerous anthropic interventions. On the other hand, the diversity of the Pantanal environments and especially its nature as a continent-sized settling basin make it the subject of a very wide array of human impacts. Alho et al. (1988) present a fairly updated and very comprehensive listing of the environmental worries.

Many projects were proposed in the past, in order to obtain maximum benefit (sic!) from the Pantanal. There has been a project to use the huge biomass of floating plants of the Pantanal as a source for bio-gas. There was a proposal of expanding rice culture and for this end to transform the

Pantanal into an immense chessboard of "polders". After the floods of the 70's hundreds of kilometres of dikes and dams, so-called "aterros" were built to protect pasture land. While their efficiency during another, future, cycle of large floods is still to be seen, they have brutally interfered with the natural migration routes of the fish fauna. Instead of high quality "piracema" fishes, low quality non-migratory fishes are promoted. The dikes also promote the phenomenon of "águas ruins" (badwaters), since with insufficient water exchange, many flooded areas develop anoxic conditions in the water, a source of massive fish mortality.

All these problems of water regulation, will bring damage at a very much higher scale, when the canalization of Rio Paraguay will be carried out.

In the terms of the "Hidrovia" project, the river will be deepened, the many meanders cut short and canalized and the inflow from the surrounding swamps will be kept to a minimum. Whereas the economic advantages of a sea-going river shipping, reaching Cáceres, have still to be evaluated, the ecological disaster will certainly be of an unprecedented magnitude.

The deforestation and large-scale agricultural development on the Planalto sends large loads of eroded sediment into the Pantanal rivers. The most endangered is Rio Taquari, once a navigable river. More than 3 million tons of sand are carried by this river (Ribeiro, 1987). Another heavily clogged river is Rio Coxim. In both of them the "piracema" is endangered.

For a land, like Brazil, where hydroelectrical plants are so numerous, the Pantanal suffers little from this problem. A dam has been built on Rio Manso, a tributary of Rio Cuiabá, in order to service a hydroelectric plant. Although the plant is not producing yet, hydraulic damage has already been done and the fish migration has been interfered with.

The rivers are also increasingly polluted with mercury. The "garimpeiros", the gold prospectors work on a fleet of many hundreds of boats, especially near Poconé and on Rio Cuiabá and Rio Bento Gomes. They use mercury to amalgamate the gold which they find in the river beds. Afterwards, they "burn" the mercury off, in order to obtain the pure gold. The mercury reaches the ecosystems mainly by air, in the form of methyl mercury and less goes directly to the water and the sediment (Esteves, 1988). However, atmospheric pollution with mercury exposes the Pantanal also to the influence of the much more intensive activity of irregular gold prospecting in the basin of Rio Madeira.

Lacerda et al. (1991) indicate the year 1981 as the background date from which onward mercury accumulation in the sediments can be detected. Although accumulation rates found by these authors for the lakes in the Poconé area range from 90 to 120 μg Hg per square metre and per year, the levels are not dangerously high as yet. Similarly Vieira (1994), studying the uptake of mercury in ampullariid snails and in snail-eating birds in the area of Rio Bento Gomes, still did not find dangerous levels, although the rates are increasing. Even in kingfishers and raptors, the levels did not much exceed

1,00 ppm (Kitayama in Alho et al., 1988). Yet Vieira (1991) mentions already an adverse influence of mercury on the reproductive behaviour of some birds. It appears that mercury pollution is a problem which might very soon reach dangerous levels within the large "settling pond" that is the Pantanal. The only other type of industrial pollution suffered by the Pantanal rivers, comes from the alcohol distilleries. Alho et al. (1988) and Coutinho et al. (1994) mention also the noxious effect of the stillage produced by the industries which turn sugar cane into fuel. These industries are in their majority situated outside the Pantanal, on the Planalto, but their effluents go to our region. Alho et al. (1989) warn that alcohol distilleries are being planned in the Pantanal proper, although the state of Mato Grosso do Sul has banned such industries from the catchment of Rio Paraguay. The stillage contains much organic matter in the form of fibrous vegetal matter, but can also lower the pH and increase the eutrophicity of the rivers carrying efluents.

Finally, the rivers of the Pantanal are often polluted by agrotoxics. Since the recent trend on the great land properties to switch to extensive soy-bean and rice growing, the danger of this sort of pollution has increased. Agriculture in the upland areas which drain towards the Pantanal is very intensive for years. Herbicides, are used Vietnam-style, in order to fight against unwanted vegetation which grows on over-pastured fields, or otherwise disturbed grasslands(see above). Alho et al. (1989) reports that the ill-famed defoliant "Agent White" is being air-sprayed and that there is also a suspicion that "Agent Orange" is being used. There is, however, no monitoring of the impact and accumulation of these herbicides in the Pantanal waters of in their fauna.

Insecticides are largely used, especially on the upland farms. Fereira et al. (1994) mention especially the organophosphates Monocrotophos and Methyl Parathion, besides some chlorates. These authors citing Yamaciro (1989, non vidi), speak of an annual official insecticide use of over 700,000 l and over 115,000 kilograms of biocides in the state of Mato Grosso do Sul alone and add that the actual, unreported use has been probably 70–90% higher than that.

To conclude: The man-induced disturbances in the Pantanal system are not being closely followed and although superficial observations seem to indicate that the system has a high degree of resilience, the whole picture is dangerous: The Pantanal is an almost endorheic system, an enormous settling pond with practically no flushing, in which a build-up of noxious substances can be expected to increase exponentially to catastrophic levels.

30. Nature conservancy and ecotourism

The Pantanal was largely unknown to the general public when in the late 80's the soap-opera "Pantanal" by the Manchete television channel brought its beauty to millions of houses. It is indeed an uncommon situation, that a large area in need of conservation steps directly, without any intermediate stage, from oblivion into the jet and television age.

As tens of thousands of tourists, both Brazilian and from abroad converge in less than one decade on the Pantanal, the basics of the environments are still largely unknown and as a consequence, a right conservation policy is still in its infancy.

The main campuses of the two federal state universities, in Cuiabá, respectively Campo Grande, are outside the Pantanal proper. The university of Mato Grosso do Sul has a filial campus in Corumbá, but the only active research on the environment of the Pantanal is done by EMBRAPA, a governmental company of agricultural research. The majority of the few researches and publications reviewed in this volume, result from the Embrapa facilities in Nhecolândia and almost all scientists actively working on Pantanal-related issues are associated with this company.

Unlike Amazonia, which has attracted countless local and foreign researchers, long before the tourists invaded, the Pantanal is still unknown by the scientific community abroad, although it figures high in the current editions of the tourist guide books.

Whereas in Amazonia international research is structured around large scientific institutions like INPA in Manaus or the Goeldi Museum in Belem, to speak of Brazil alone, there is no comparable scientific institution in the Pantanal. Needless to say that an international research effort in the Pantanal is already long overdue.

The 1988 constitution of Brazil proclaimed the Pantanal as a " National Patrimony" of Brazil, but there is today no nature reserve in the Pantanal in the accepted sense of the word. The river island of Taiamá in the northern Pantanal was set aside as a "National Park", also called "Parque Nacional do Caracará"in the early years of 1970 (Jorge Padua and Coimbra Filho, 1979) (Fig. 4). The surface of this park is of only 145,000 hectares and according to specialists is totally non-representative for the Pantanal, since 90% of it is under water during the rainy season. Anyhow, according to Alho (1986), the park is grievously understaffed and practically abandoned. According to Gonçalves and Bluma (in Alho et al., 1989), the ideal site for a representative conservational area would be the Serra do Amolar, where all the environments of the Pantanal are represented from the dry mountain flora to the large lake of Gaiba. Recently an international Pantanal reserve is being roughly drafted in the area between Coimbra and Bahia Negra, where Brazil, Bolivia and Paraguay meet in a triple frontier.

Brazil has excellent laws of nature conservation and a very active Federal Agency, the IBAMA. But the "Policia Florestal", and the very agents of Ibama are disproportionately few and underpaid. In 1987, there were less than 20 such agents for the whole Pantanal (Ribeiro, in "Globo Rural" 2/16 January 1987). The forest rangers received an airplane in 1989, but it came to be lent for the use of local politicians. The control on the main access roads is so leaky that according to Emiko Kawakami (U. Capozoli "O Estado de São Paulo", 12 April 1993), some 8 million tons of fish are smuggled out of the Pantanal yearly.

In a region in which the federal authorities are far away, the law of the land owners is still almost unchallenged.

Strange as it seems, but all the "land" of the Pantanal is carved up between the different fazendeiros. By law only the banks of the navigable rivers are Federal property. This is why Dr. Paulo Nogueira Neto of the University of São Paulo and once Federal Secretary for Environment, suggested as a last recourse to call in the Brazilian navy to patrol the rivers of the Pantanal.

However, the combined action of education and of the novel sources of profit are at work even in the Pantanal. Many land-owners turned into passionate nature protectors.

Near the confluence of the Itiquira and the Piquiri, the 120,000 hectares of the Lutz family contain now a very important wildlife sanctuary. Near Miranda, fazenda Caiman of the Klabin family, is a "refúgio", in which 7,000 hectares of a total of 53,000 is a strictly preserved wilderness.

More and more properties, perhaps already some 100 all over the Pantanal are turning into "hoteis fazenda" or "pousadas", serviced by a fleet of four-wheel cars, speed boats and often small planes, which offer to the visitors a rich programme of safari trails and "pantaneiro" lore. To stimulate this profitable trade, a diversity of Pantanal animals surround the visitor even on the very premises of the hotels. There are "ecological " fazendas where the henchmen of the landowners, once armed guards against cattle thieves and poachers, turned into efficient rangers. They are often so dedicated and so fierce in protecting the natural objects as they were in their past avatars.

The "non-governmental" nature conservation entities, like the Sodepan (Sociedade da Defesa do Pantanal), keep a wary eye on this new developments. The disembarking tourist groups can be a destructive force, like when they disturb and frighten the "ninhais", the bird rookeries. But the experience of Brazil is such, that the so-called ecotourism, with all its shortcomings, represents a powerful factor. It brings to the Pantanal scores of unpaid (and paying) nature wardens and creates powerful pressure groups all over the country. For the pantaneiros, from the rich landowner to the destitute migrant farm hand, the ecotourism is an increasing bonanza for new revenues.

Like in a huge endorheic lake, the build up of the adverse inflows and the impact of the irresponsible engineering projects will take some time to show its effects. Since in the words of Arnildo Pott from Embrapa Corumbá, the Pantanal ecosystems are constituted by the most rugged species of the surrounding regions, the effects will be slow to appear. There will probably be no species which will disappear altogether for science: the populations are too big and the land is open. But the large and colourful populations which make the pride of this unique environment of the globe will dwindle and we all will have lost one of the most beautiful showcases of plant and animal life on Earth.

References

Ab'Saber, A.N. 1988. O Pantanal Mato-Grossense e a teoria dos refúgios. *Revista Brasileira de Geografia* 50, special, 1–2: 9–57.
Ab'Saber, A.N. 1989. Debate. In: Ed. Sayd de Carvalho, S. *Anais do I Simpósio Internacional sobre Conservação do Pantanal.* SEMA MS-WWF, Campo Grande 108–112.
Adamoli, J.A. 1981. O Pantanal e suas relações fitogeográficas com o cerrado. Discussão sobre o conceito "Complexo do Pantanal", Anais da Sociedade Brasileira de Botânica. *Congresso Nacional de Botânica 32, Teresina* 109–119.
Adamoli, J.A. 1986a. A Dinâmica da Inundações no Pantanal. In: Ed.: Boock, A. *Anais do 1 Simpósio sobre Recursos Naturais e Sócio-Econômicos do Pantanal*, pp. 51–62. EMBRAPA-CPAP Documentos 5, Corumbá, MS.
Adamoli, J.A. 1986b. Fitogeografia do Pantanal. In: Ed. Boock, A. *Anais do 1 Simpósio sobre Recursos Naturais e Sócio-Econômicos do Pantanal*, pp. 105–106. EMBRAPA-CPAP Documento 5, Corumbá, MS.
Alfonsi, R.R. and Paes de Camargo, M.B. 1986. Condições Climáticas para a Região do Pantanal Mato-Grossense. In: Ed. Boock, A. *Anais do 1 Simpósio sobre Recursos Naturais e Sócio-Econômicos do Pantanal*, pp. 29–42. EMBRAPA-CPAP Documentos 5, Corumbá, MS.
Alho, C.J.R. 1986. Manejo da Fauna Silvestre. In: Ed. Boock, A. *Anais do 1 Simpósio sobre Recursos Naturais e Sócio-Econômicos do Pantanal*, pp. 183–197. EMBRAPA-CPAP Documento 5, Corumbá, MS.
Alho, C.J.R., Lacher, T.E. and Gonçalves, H.C. 1988. Environmental Degradation in the Pantanal Ecosystem. *Bioscience* 38(3): 164–171.
Almeida, F.F.M. de 1945. Geologia do Sudoeste Matogrossense. *Boletim da Divisão de Geologia e Mineralogia* 116.
Almeida, F.F.M. de 1959. O Planalto Centro Occidental e o Pantanal Matogrossense. Guia de excursão. *União Internacional de Geografia Rio de Janeiro* 3: 169.
Almeida, F.F.M. de and Lima, M.A. de 1956. Excursion Guidebook. Vol. 1. *18th International Geography Congress*, Rio de Janeiro.
Anonymous 1982. *Mapa Geológico* 1:1.000.000 Corumbá Projeto Radambrasil.
Arruda Mauro, R. de, Mourão, G.M. de, Silva, M.P. da, Coutinho, M.E., Moraes Tomas. W. and Magnusson, W.E. 1994. Quantos Cervos Existem no Pantanal? *XX Congresso Brasileiro de Zoologia, Resumos*, pp. 121–122.
Ashton, P.J. and Mitchell, D.S. 1989. Aquatic Plants: Patterns and Modes of Invation, Attributes of Invading Species and Assessment of Control Programmes, pp. 111–154. In: Eds. Drake, J.A., Mooney, H.A., di Castri, F., Groves, R.H., Kruger, F.J., Rejmanek, M. and Williamson, M. *Biological Invasions: a Global Perspective.* John Wiley & Sons, Chichester.
Bánárescu, M. 1990. *Zoogeography of Fresh Waters.* Vol. 1. Aula, Wiesbaden.
Barthold, P. 1993. *Bird Migration. A General Survey.* Oxford University Press, Oxford.
Beck, S.G. 1984. Comunidades vegetales de las sabanas inundadizas en el NE de Bolivia. *Phytocoenologia* 12: 321–350.
Berthold. T. 1991. Vergleichende Anatomie, Phylogenie und Historische Biogeographie der *Ampullariidae* (Mollusca, Gastropoda). *Abhandlungen des Naturwissenschaftlichen Vereins in Hamburg* (NF)29:256.
Bonetto, A.A. 1975. Hydrologic Regime of the Paraná River and its Influence on Ecosystems. In: Ed. Hassler, D. *Coupling of Land and Water Systems*, pp. 175–197. Springer Verlag.
Bonetto, A.A. and Wais, I.R. 1990. The Paraná river in the framework of modern paradigms of fluvial systems. *Limnologia Brasileira* 3: 164–180.
Braun, E.H.G. 1977. O cone aluvial do Taquari, unidade geomórfica marcante na planice quaternária do Pantanal. *Revista Brasileira de Geografia* 39(4):164–180.
Brown, K.S.J. 1986. Zoogeografia da Região do Pantanal Mato-Grossense. In: Ed. Boock, A. *Anais do 1 Simpósio sobre Recursos Naturais e Sócio-Econômicos do Pantanal*, pp. 137–178. EMBRAPA-CPAP Documento 5, Corumbá, MS.

Brum, P.A.R. de and Sousa, J.C. de 1985. Níveis de nutrientes minerais para gado, em lagoas ("baías" e "salinas") no Pantanal Sul Mato-Grossense. *Pesquisa Agropecuária brasileira* 20(12): 1451–1454.
Cadavid-Garcia, E.A. 1992. *Estado do conhecimento sócio-econômico e técnico-científico da Bacia do Alto Paraguai com enfase no Pantanal Mato-Grossense: Resumo Informativo.* Technical Report. Embrapa-CPAP, Corumbá.
Camargo, F.C. de 1968. Agricultura na América do Sul. In: Eds. Fittkau, E.J., Illies, J., Klinge, H., Schwabe, G.H. & Sioli, H. *Biogeography and Ecology in South America.* Vol. 1, pp. 302–328. Dr. W. Junk, The Hague.
Campos, F.V. 1969. *Retrato de Mato Grosso.* Brasil Oeste Editora Ltda., São Paulo.
Cardoso da Silva, T. 1986. Contribuição da Geomorfologia para o Conhecimento e Valorização do Pantanal. In: Ed. Boock, A. *Anais do 1 Simpósio sobre Recursos Naturais e Sócio-Econômicos do Pantanal,* pp. 77–90. EMBRAPA-CPAP Documentos 5, Corumbá, MS.
Carter, G.S. and Beadle, B.A. 1929. The Fauna of the Swamps of the Paraguayan Chaco in relation to its Environment. I. Physico-Chemical Nature of the Environment. *Linnean J. Zoology* 37: 205–257.
Carvalho, J.C. 1986. Fauna Terrestre e Aquática. In: Ed. Boock, *Anais do 1 Simpósio sobre Recursos Naturais e Sócio-Econômicos do Pantanal.* pp. 179–182. EMBRAPA-CPAP Documento 5, Corumbá, MS.
Cei, J.M. 1980. *Amphibians of Argentina.* Monitore Zoologico Italiano N.S. Monografia 2.
Correa Filho, V. 1946. *Pantanais matogrossenses (Desenvolvimento e ocupação).* Biblioteca Geografica Brasileira, Rio de Janeiro.
Costa, C.A.F. and Cato, J.B. 1994. Helminth parasites of capybaras (*Hydrochoerus hydrochaeris*) in the sub-region of Nhecolândia, Pantanal, state of Mato Grosso do Sul. *Revista Brasileira de Biologia* 54(1): 39–48.
Coutinho, L.M. 1990. Fire in the Ecology of the Brazilian Cerrado. In: Ed. Goldammer, J.G. Fire in the Tropical Biota, pp. 82–105. Springer Verlag, Berlin-Heidelberg.
Coutinho, M.E., Mourão, M.G., Pereira Silva, M. and Campos, Z. 1994. The sustainable use of natural resources and the conservation of the Pantanal wetlands, Brazil. *Acta Limnologica Brasiliensia* 5: 165–176.
Crawshaw, P.G., Jr. 1991. Effects of Hunting on the Reproduction of the Paraguayan Caiman (*Caiman yacare*) in the Pantanal of Mato Grosso, Brazil. In: Eds. Robinson, J.G. and Redford, K.H. *Neotropical Wildlife Use and Conservation,* pp. 145–154. University of Chicago Press, Chicago.
Cunha, J. 1943. Análise química das águas. Cobre do Jauru: lagoas alcalinas do Pantanal. *Boletim do Laboratório de Produção Mineira* 6:18–19.
Davino, A. 1968. Determinação das espessuras dos sedimentos do Pantanal Mato-Grossense por sondagens elétricas. *Anais da Academia Brasileira de Ciências* 40(3):327–330.
Diegues, A.C.S. Ed. 1990. *Inventário das áreas úmidas do Brasil.* USP-IUCN-Ford Foundation.
Dioni, W. 1967. Investigacion preliminar de la estructura basica de las asociaciones de la micro e meso fauna de las raizes de las plantas flotantes. *Acta Zoologica Lilloana, Tucuman* 23:111–132.
Dourojeanni, M.J. 1980. *Desarrollo y conservacion en el Pantanal Matogrossense (Brasil) con especial referencia al manejo de la fauna silvestre.* Report. Instituto Brasileiro de Desenvolvimento Florestal IBDF Brasilia.
Dubs, B. 1992. *Birds of Southwestern Brasil.* Betrona Verlag, Kusnacht, Switzerland.
Eiten, G. 1983. *Classificação da Vegetação do Brasil.* CNPq/Coordenação Editorial, Brasilia.
Esteves, F. 1988. *Fundamentos da Limnologia.* Interciências, Rio de Janeiro.
Fairchild, T.R. 1978. Evidências paleontológicas de uma possível idade "Ediacarana" ou Cambriana inferior, para parte do Grupo Corumbá (Mato Grosso do Sul). Boletim da Sociedade Brasileira de Geologia, *XXX Congresso Brasileiro de Geologia*: 181.

Fereira, C.J.A., Soriano, B.M.A., Galdino, S. and Hamilton, S.K. 1994. Anthropogenic factors affecting waters of the Pantanal wetland and associated rivers in the Upper Paraguay river basin of Brazil. *Acta Limnologica Brasiliensia* 5: 135–148.

Fereira, J.V.C. 1994. Pantanal – Tipos e Aspectos do Brasil. *Revista Brasileira de Geografia* 6(2):281–285.

Ferraz de Lima, J.A. 1981. A pesca no Pantanal do Mato Grosso (Rio Cuiabá): Biologia e ecologia pesqueira. *Anais do II Congresso Brasileiro de Engenharia de Pesca*: 503–516.

Ferraz de Lima, J.A. 1986/87. A pesca do Pantanal do Mato Grosso (Rio Cuiába): Importância dos peixes migratórios. *Acta Amazonica* 16/17:87–94.

Figueredo Alvarenga, L.C. de 1981. Bivalvia. In: Eds. Hurlbert, S.H., Rodriguez, G. and Dos Santos, N.D. *Aquatic Biota of Tropical South America*. Vol. 2 *Anarthropoda*, pp. 207–211. San Diego State University, San Diego.

Fittkau, E.J. 1969. The Fauna of South America – In: Eds. Fittkau, E.J., Illies, J., Klinge, H., Schwabe, G.H. and Sioli, H. *Biogeography and Ecology in South America*. Vol. 2, pp. 624–658. Dr. W. Junk, The Hague.

Fukuhara, H. and Henry, R. 1987. Standing crop of zoobenthos in a small lake in Northern Pantanal. In: Eds. Saijo, Y. and Tundisi, J.G. *Limnological Studies in Rio Doce Valley Lakes and Pantanal Wetland, Brazil*, pp. 183–185. Water Research Institute Nagoya University, Nagoya.

Fukuhara, H. and Mitamura, O. 1985. Standing crop of zoobenthos in the Pantanal. In: Eds. Saij, Y. and Tundisi, J.G., *Limnological Studies in Central Brazil*, pp. 197–201. Water Research Institute Nagoya, Nagoya.

Gery, J. 1969. The Fresh-Water Fishes of South America. In: Eds. Fittkau, E.J., Illie, J., Klinge, H., Schwabe, G.H. and Sioli, H. *Biogeography and Ecology in South America*. Vol. 2, pp. 828–848. Dr. W. Junk, The Hague.

Gery, J. 1977. *Characoids of the World*. T.F.H. Publication, Neptune, New Jersey.

Godoi Filho, J.D. 1986. Aspectos Geológicos do Pantanal Mato-Grossense e de sua Área de Influência. In: Ed. Boock, A. *Anais do 1 Simpósio sobre Recursos Naturais e Sócio-Econômicos do Pantanal*, pp. 63–76. EMBARAPA-CPAP Documentos 5, Corumbá, MS.

Goldstein, R.J. 1973. *Cichlids of the World*. T.F.H. Publishers, Neptune, New Jersey.

Gomes de Souza, L. 1986. Retrospectiva Histórica do Pantanal. In: Ed. Boock, A. *Anais do 1 Simpósio sobre Recursos Naturais e Sócio-Econômicos do Pantanal*, pp. 199–203. EMBRAPA-CPAP Documento 5, Corumbá, MS.

Goodland, R.J.A. 1971. A physiognomic analysis of the "Cerrado" vegetation of Central Brasil. *Journal of Ecology* 59:411–429.

Goulding, M. 1980. *The fishes and the forest: explorations in Amazonian natural history*. University of California Press, Berkeley.

Goulding, M. 1981. *Man and Fisheries on an Amazon Frontier*. Dr. W. Junk, The Hague.

Goulding, M., Carvalho, M.L. and Fereira, E.G. 1988. *Rio Negro, Rich Life in Poor Waters*. SPB Academic Publishers, The Hague.

Grabert, H. 1967. Sobre o desaguamento natural do sistema fluvial do Rio Madeira desde a construção dos Andes. *Atas do Simpósio sobre Biota Amazonica* 1:209–214.

Haase, R. and Beck, S.G. 1989. *Structure and composition of savanna vegetation in northern Bolivia: A preliminary report*. Brittonia 41:81–100.

Harry, H.W. 1962. A critical catalogue of the nominal genera and species of *Neotropical Planorbidae. Malacologia* 1(1):33–53.

Heckman, C.W. 1994. New limnological nomenclature to describe ecosystem structure in the tropical wet-and-dry climatic zone. *Archiv für Hydrobiologie. Ergebnisse der Limnologie* 130:385–407.

Heckman, C.W., Lopes Hardoim, E., Ferreira, S.A. and Kreztschmar, A.U. 1993. Preliminary Observations on Some Cosmopolitan Algae in Ephemeral Water Bodies of the Pantanal, Mato Grosso, Brazil. In: Eds. Gopal, H., Hillbricht-Ilkowska, A. and Wetzel, R.G. *Wetlands*

and Ecotones: Studies on Land-Water Interactions, pp. 279–292. National Institute of Ecology, New Delhi.
Hoehne, F.C. 1936. O Grande Pantanal de Mato Grosso. *Boletim de Agricultura*, São Paulo: 443–447.
Howell, P., Lock, M. and Cobb, S. 1988. *The Jonglei Canal*. Cambridge University Press, Cambridge.
Jaeckel, S.C.A.J. 1969. Die Mollusken Sudamerikas. In: Eds. Fittkau, E.J., Illies, J., Klinge, H., Schwabe, G.H. and Siolo, H. *Biogeography and Ecology of South America* Vol. 2, pp. 794–827. Dr. W. Junk, The Hague.
Janssen, A. 1986. Flora und Vegetation der Savannen von Humaitá und Standortsbedingungen. *Disesrtationes Botanicae* 93:1–321.
Jedicke, A., Furch, B., Saint-Paul, U. and Brigitte Schluter, J. 1989. Increase in the oxygen concentration in Amazon waters resulting from the root exudation of two notorious water plants *Eichhornia crassipes (Pontederiaceae)* and *Pistia stratiotes (Araceae)*. *Amazoniana* 11(1):53–69.
John, D.M. 1986. The Inland Waters of Tropical West Africa. Archiv für Hydrobiologie. *Ergebnisse de Limnologie* 23:1–244.
Junk, W.J. 1970. Investigation in the Ecology and Production-Biology of the "Floating Meadows" (Paspalo-Echinochloetum) of the Middle Amazon. Part 1. *Amazoniana* 2(4):449–495.
Junk, W.J. 1973. Investigations on the Ecology and Production-Biology of the "Floating Meadows" (Paspalo-Echinocloetum) of the Middle Amazon. Part II. The Aquatic Fauna of the Root Zone of Floating Vegetation. *Amazoniana* 4(1):9–102.
Junk, W.J. 1993. Wetlands of Tropical South America. In: Eds. Whigham, D.F., Dykyjova, D. and Hejny, S. *Wetlands of the World: Inventory, Ecology and Management*. Vol. 1. pp. 679–739. Kluwer Academic Publishers, Dordrecht.
Kawakami de Resende, E. 1989. Recursos naturais do Pantanal. *Anais do I Congresso Internacional sobre Conservação do Pantanal*, pp. 26–29. Campo Grande.
Killeen, T.J. and Hinz, P.N. 1992. Grasses of the Precambrian shield region in Eastern lowland Bolivia: I. Habitat preferences. *J. Tropical Ecology* 8(4):389–407.
Klammer, G. 1982. Die Palaeowuste des Pantanal von Mato Grosso und die pleistozane Klimageschichte der brasilianischen Randtropen. *Zeitschrift für Geomorphologie* 26(4):393–416.
Komissarov, B.N. 1988. *A Expedição do Acadêmico G.I. Langsdorff e seus Artistas ao Brasil*. Ed. Alumbramentos, Rio de Janeiro.
Kretzschmar, A., Ferreira, S.A., Lopes Hardoim, F. and Heckman, C.W. 1993. Peak Growth of the *Asplanchna sieboldi* (Leydig, 1854) Rotifer Aggregation in Relation to the Seasonal Wet and Dry Cycle in the Pantanal, Mato Grosso, Brazil. In: Eds. Gopal, B., Hillbricht-Ilkowska, A. and Wetzel, R.G., *Wetlands and Ecotones: Studies on Land-Water Interactions*, pp. 293–301. National Institute of Ecology, New Delhi.
Krieg, H. 1936. Tiegeographische Wirkungen der jahrlichen Uberschwemmungen im Stromgebiete des Paraguay-Paraná. *Isis* 1934–1936:5–7.
Lacerda, L.D.D., Salomons, W., Pfeiffer, W.C. and Bastos, W.R. 1991. Mercury distribution in sediment profiles from lakes in the high Pantanal, Mato Grosso State, Brazil. *Biogeochemistry* 14(2):91–98.
Lamonica-Freire, E.M. de, Bicudo, C.E.D.M. and De Castro, A.A.J. 1992. Algal flora of the Pocone Pantanal, State of Mato Grosso, Brazil. I. Euglenaceae. *Revista Brasileira de Biologia* 52(1):141–149.
Levi-Strauss, C. 1976. *Tristes Tropiques*. Penguin Books.
Levi-Strauss, C. 1994. *Saudades do Brasil*. Companha das Letras, São Paulo.
Lobato Paraense, W. 1981. Gastropoda. In: Eds.: Hurlbert, S.H., Rodriguez, G. and Dos Santos, N.D. *Aquatic Biota of Tropical South America*. Vol. 2. *Anarthropoda*, pp. 191–196. San Diego State University, San Diego.
Lowe-McConnel, R.H. 1975. *Fish communities in tropical fresh water*. Longman, London.

Lowe-McConnell, R.H. 1987. *Ecological Studies in Tropical Fish Communities.* Cambridge University Press, Cambridge.
MacDonald, D.W. 1981. Dwindling resources and the social behaviour of Capybaras (*Hydrochoreus hydrochaeris*) (Mammalia). *J. Zoology London* 194:371–391.
Magalhães, N.W. de 1992. *Conheça o Pantanal.* Terragraph Artes e Informática, São Paulo.
Marins, R.V., Conceiçao, P.N. and Lima, J.A.F. 1981. *Estudos ecológicos das principais espécies de peixes de interesse comercial, esportivo e ornamental da Bacia do Alto, Paraguai.* Ed. SEMA, Brasília.
Marques, E.J. and Uetanabaro, M. 1993. Ninhais do Rio Vermelho, Pantanal do Abobral, MS: Um Estudo de Caso, Dados Preliminares. *Congresso Brasileiro de Ornitologia II*, pp. 10–11. São Leopoldo.
Medem, F. 1983. *Los Crocodylia de Sur America.* Vol. 2. Universidad Nacional de Colombia, Bogotá.
Menezes, N.A. 1974. Redescription of the genus Roestes (*Pisces, Characidae*). *Papéis Avulsos de Zoologia, São Paulo* 27(17):219–225.
Mitamura, O., Hino, K., Saijo, Y., Tundisi, J.G., Matsumura-Tundisi, T., Ikusima, I., Sunaga, T., Nakamoto, N. and Fukuhara, H. 1985. Physico-chemical features of the Pantanal water system. In: Eds. Saijo, Y. and Tundisi, J.G. *Limnological Studies in Central Brazil*, pp. 189–196. Water Research Institute Nagoya, Nagoya.
Moraes Paula, A. and Aguiar Ferreira, S. 1994. Levantamento Preliminar de Espécies de Gafanhotos Semiaquáticos (*Orthoptera, Acridoidea*) e suas Macrófitas Hospedeiras do Pantanal de Pocone, MT. *Resumos do XX Congresso Brasileiro de Zoologia*, p. 41. Rio de Janeiro.
Morelli, J.H. and Adamoli, J.A. 1973. Subregiones ecologicas de la provincia del Chaco. *Ecologia* 1(1):29–33.
Mourão, G.D. de 1989. *Limnologia Comparativa de Três Lagoas (Duas "Baías" e uma "Salina") do Pantanal da Nhecolandia, MS.* MSc Thesis, Universidade Federal de São Carlos.
Mourão, G.M. de, Ishii, T.H. and Campos, Z.M.S. 1988. Alguns fatores limnológicos relacionados com a ictiofauna de baías e salinas do Pantanal da Nhecolândia, Mato Grosso do Sul, Brasil. *Acta Limnologica Brasiliensia* 2:181–198.
Neiff, J.J. 1990. Aspects of primary productivity in the Lower Paraná and Paraguay rives. *Acta Limnologica Brasilensia* 3:77–113.
Neves, L.D., Isaias, R.M.D.S. and Mello Filho, L.E.D. 1993. Anatomic study of the leaf of *Ficus elliotiana* Moore. *Bradea* 6(22):196–205.
Noodt, W. 1969. Die Grundwasserfauna Südamerikas. In: Eds. Fittkau, E.J., Illies, J., Klinge, H., Schwabe, G.H. and Sioli, H. *Biogeography and Ecology in South America.* Vol. 2, pp. 659–684. Dr. W. Junk, The Hague.
Oliveira Carvalho, A.E. 1986. Hidrologia da Bacia do Alto Paraguaj. In: Ed. Boock, A. *Anais do 1 Simpósio sobre Recursos Naturais e Sócio-Econômicos do Pantanal*, pp. 43–50. EmBRAPA-CPAP-UFMS. Documentos 5, Corumbá, MS.
Oliveira Filho, A.T. de 1992. The vegetation of the Brazilian "murundus": The island-effect on the plant community. *J. Tropical Ecology* 8(4): 465–486.
Padua, J.M.T. and Coimbra, A.F. 1979. *Os Parques Nacionais do Brasil.* INCAFO, Madrid.
Pacheco do Amaral Filho, Z. 1986. Solos do Pantanal Mato-Grossense. In: Ed. Boock, A. *Anais do 1 Simpósio sobre Recursos Naturais e Sócio-Econômicos do Pantanal*, pp. 91–103. EMBARPA-CPAP Documento 5, Corumbá, MS.
Padua Bertelli, A. de 1988. *O Pantanal Mar dos Xaráes.* Edições Siciliano, São Paulo.
Parodiz, J.J. and Bonetto, A.A. 1963. Taxonomy and zoogeographic relationships of South American Naiades (*Pelecypoda: Unionacea and Mutelacea*). *Malacologia* 1(2):179–213.
Peracca, M.G. conte de 1904. Viaggio del Dr. Borelli nel Matto Grosso Brasiliano e nel Paraguay, 1899. Rettili ed Anfibi. *Boletino Museo Anatomia Comparata Universita de Torino* 19 (460):1–15.

Petri, S. and Fulfaro, V.J. 1988. *Geologia do Brasil.* T.A. Queiroz and Editora USP, São Paulo.
Pfaffstetter, O. 1974. A swampland model. In: *Mathematical models in hydrology.* Vol. 2, pp. 476–482. International Association of Hydrology-UNESCO-WMO.
Pinto Braga, M. 1988. *Langsdorff de Volta. Introduction to a Catalogue of drawings and aquarels by Rugendas, Taunay and Florence.* Fundação Pro-Leitura, Rio de Janeiro.
Pinto Paiva, M. 1984. *Aproveitamento de recursos faunisticos do Pantanal de Mato Grosso: Pesquisas necessárias e desenvolvimento de sistemas de produção mais adequados a região.* EMBRAPA Depto. de Difusão Técnica, Brasilia.
Poi de Neiff, A. and Neiff, J.J. 1984. Dinámica de la vegetación acuatica flotante y su fauna en charcos temporarios del sudeste del Chaco (Argentina). *Physis B* 42:103:53–67.
Por, F.D. 1992 *Sooretama – The Atlantic Rain Forest of Brazil.* SPB Academic Publishing, The Hague.
Prado, D.E. 1993. What is the Gran Chaco vegetation of South America? II. A redefinition. Contribution to the study of the flora and vegetation of the Chaco. VII. *Candollea* 48(2):615–629.
Prance, G.T. and Schaller, G.B. 1982. Preliminary study of some vegetation types of the Pantanal, Mato Grosso, Brazil. *Brittonia* 14(2):15–46.
Putzer, H. 1984. The geological evolution of the Amazon basin and its mineral resources. In: Ed. Sioli, H. *The Amazon: Limnology and landscape ecology of a mighty tropical river and its basin*, pp. 15–46. Dr. W. Junk, The Hague.
Reclus, E. 1895. *The Earth and its Inhabitants. South America. 2. Amazonia and La Plata.* D. Appleton, New York.
Reid, J.W. and Moreno, I.H. 1990. The Copepoda (Crustacea) of the southern Pantanal, Brazil. *Acta Limnologica Brasiliensia* 3:721–739.
Resende, E.K. 1989. Recursos Naturais do Pantanal. In: Ed. Sayd de Carvalho, S. *Anais do 1 Congresso Internacional sobre Conservação do Pantanal*, pp. 26–29. SEMA MS; WWF, Campo Grande.
Ribeiro, J.H. 1987. Tuiuiu, o primeiro grito de socorro. *Globo Rural Janeiro* 2(16):62–79.
Saijo, Y., Mitamura, O., Ikusima, I., Tundisi, J.G., Sunaga, T., Nakamoto, N., Fukuhara, H., Barbosa, F.A.R., Henry, R., Seixas Filho, C.T. and Silva, V.P. 1987. Physico-Chemical Features of Small Lakes near Porto Jofre in Northern Pantanal. In: Eds. Saijo, Y. and Tundisi, J.G. *Studies in Rio Doce Lakes and Pantanal Wetland, Brazil*, Vol. 2, pp. 177–182. Water Research Institute Nagoya University, Nagoya.
Saint Paul, U. 1984. Physiological adaptation to hypoxia of a neotropical characoid fish, Collosoma macropomum, Serrasalmidae. *Environ. Biol. of Fishes* 11:1:53–62.
Santos, S.A. and Nogueira, M.J.S. 1994. Identificação da Dieta de Caiman crocodilus yacare em Diversos Ambientes do Pantanal Mato-Grossense. *Resumos do XX Congresso Brasileiro de Zoologia*, p. 107. Rio de Janeiro.
Sazima, I. 1988. Territorial behavior of scale-eating and herbivorous Neotropical characiform fish. *Revista Brasileira de Biologia* 48(2):189–194.
Sazima, I. and Zamprogno, C. 1985. Use of water hyacinths as shelter, foraging place, and transport by young piranhas, Serrasalmus spilopleura. *Environ. Biol. of Fishes* 12:3:237–240.
Schaller, G.B. and Crawshaw, P.G. 1982. Fishing behavior of the Paraguayan caiman (*Caiman crocodilus*). *Copeia* 66–72.
Schaller, G.B. and Vasconcelos, J.M.C. 1978. A marsh deer census in Brazil. *Oryx* 14(4):345–351.
Schaller, G.B. and Vasconcelos, J.M.C. 1978. Jaguar predation on capybara. *Z. für Saugetierkunde* 43:296–301.
Sick, H. 1985. *Ornitologia Brasileira.* Vols. I and II. Ed. Univ. Brasilia, Brasilia.
Silva, C.J. da 1990. *Influéncia da variação do nível d'água sobre a estrutura e funcionamento de uma área alagável do Pantanal Matogrossense (Pantanal de Barão de Melgaço, MT).* Ph.D. Dissertation, Universidade Federal de São Carlos, SP.

Silva, C.J. da and Fernandes Silva, J.A. 1992. Estratégias de sobrevivência de comunidades tradicionais no Pantanal Matogrossense: Relatório Preliminar. *Nucleo de Apoio a Pesquisa sobre Populações Humanas e Áreas úmidas Brasileiras*. Estudo de Caso No. 5, São Paulo.

Silva, C.J. da and Pinto-Silva, V. 1989. Macrófitas aquáticas e as condições físicas e químicas dos "alagados", "corixos" e rios, ao longo da rodovia Transpantaneira, Pantanal Matogrossense (Poconé, MT). *Revista Brasileira de Biologia* 49(3): 691–697.

Stefan, E.R. 1964. Pantanal Matogrossense. *Revista Brasileira de Geografia* 26(3):465–478.

Sucksdorff, A. 1985. *Pantanal um paraíso perdido?* Ed. Siciliano, São Paulo.

Sucksdorff, A. 1989. Desenvolvimento e Preservação. II. Visao Nacional. *Anais do I Congresso Internacional sobre Conservação do Pantanal*. pp. 113–115. Campo Grande.

Tarifa, J.R. 1986. Sistema Climático do Pantanal: da Compreensão do Sistema a Definição de Prioridades de Pesquisa Climatológica. In: Ed. Boock, A. *Anais do 1 Simpósio sobre Recursos Naturais e Sócio-Econômicos do Pantanal*, pp. 9–28. EMBRAPA-CPAP, Corumbá MS.

Tricart, J. 1982. El Pantanal: Un ejemplo del impacto de la Geomorfologia sobre el medio ambiente. *Geografia* 7(13–14):37–50.

Tundisi, J.G. and Matsumura-Tundisi, T. 1985. The "Pantanal" Wetlands of Western Brazil. In: Eds. Saijo, Y. and Tundisi, J.G. *Limnological Studies in Central Brazil*, pp. 177–188. Water Research Institute Nagoya University, Nagoya.

Turner, P.N. and Da Silva, C. 1992. Littoral Rotifers from the State of Mato Grosso, Brazil. *Stud. Neotropical Fauna Environment* 27(4):227–241.

Updike, J. 1994. *Brazil. A Novel*. Fawcet Crest, New York.

Urban, I. 1902. Vitae itineraque collectorum botanicorum. In: Ed. Martius, K.F.P. vaon *Flora Brasiliensis*. Vol. 1, pp. 1–152.

Valverde, D. 1972. Fundamentos geográficos de planejamento rural do Município de Corumbá. *Revista Brasileira de Geografia* 34(1):49–144.

Vanzolini, P.E. and Valença, J. 1966. The genus Dracaena, with a brief consideration of Macroteiid relationships (*Sauria: Teiidae*). *Arquivos de Zoologia* 13:7–35.

Veloso, H.P. 1947. Considerações gerais sobre a vegetação do Estado de Mato Grosso. II. Notas preliminares sobre o Pantanaal e zonas de transição. *Memória Instituto Oswaldo Cruz* 45:253–272.

Veloso, H.P. 1972. Aspectos Fitoecológicos da Bacia do Alto Rio Paraguai. *Biogeografia* 7:1–31.

Vieira, L.M. 1991. *Avaliação dos níveis de mercúrio na cadeia trófica como indicador de sua biomagnifição em ambientes aquáticos da região do Pantanal*. MSc Thesis, University of São Carlos.

Vieira, L.M., Alho, C.J.R. and Ferreira, G.A. 1994. Contaminação por Mercúrio em sedimento e moluscos no Pantanal (MT). *Resumos do XX Congresso Brasileiro de Zoologia*, Rio de Janeiro.

Vieira de Campos, F. 1969. *Retrato de Mato Grosso*. Brasil Oeste, São Paulo.

Walter, H. and Breckle, S.W. 1984. Spezielle Ökologie der Tropischen und Subtropischen Zonen. *Ökologie der Erde*. Vol. 2. Gustav Fischer, Stuttgart.

Wilhelmy, H. 1958. Umlaufseen und Dammuferseen tropischer Tieflandflüsse. *Z. für Geomorphologie, NF* 2:27–54.

Wilhelmy, M. 1958. Das Grosse Pantanal. *Die Weltumschau* 18:555–559.

Willis, E.O. 1976. Effects of a cold wave on an Amazonian avifauna in the upper Paraguay drainage, western Mato Grosso, and suggestions on oscine-suboscine relationships. *Acta Amazonica* 6: 379–394.

Yamashita, C. and Vale, M. de P. 1990. Sobre ninhais de aves do Pantanal do Município de Poconé, Mato Grosso, Brasil. *Vida Silvestre Neotropical* 2(2):59–63.

Zerries, O. 1968. The South American Indians and their Culture. In: Eds. Fittkau, E.J., Illies, J., Klinge, H., Schwabe, G.H. and Sioli, H. *Biogeography and Ecology in South America* Vol. 1, pp. 329–388. Dr. W. Junk, The Hague.

Taxonomic Glossary*

Acacia sp. 50
Acari 63
Achirus 67
Acrocomia sclerocarpa (Bocaiuva)
Actinosoma pentacanthum 48
Acuri = Attaleia sp.
Aelosoma spp. 62
Aequidens paraguayensis 70, 75
Aeschynomene fluminensis 54
Aguapé = Eichhornia
Águia pesqueira = Pandion haliaetus
Ajaia ajaja (Colhereiro) 79, 81
Alface d'agua = Pistia stratiotes
Algodão bravo = Ipomoea fistulosa
Alonella sp. 62
Alouatta caraya (Bugío preto) 81, 88
Ameiva ameiva (Calango verde) 76
Amendoim do campo = Arachis glabrata
Amphipoda 63
Ampullaria 64
Ampullariidae 64
Anodorhynchus hyazinthinus (Arara azul) 83
Anas leucophrys (Marreca de coleira) 81
Ancylidae 64
Andorinha do rio = Tachycineta albiventer
Angicó = Piptadenia macrocarpa
Anhima cornuta 81
Anhinga anhinga (Biguá tinga) 77, 81
Anodontides iheringi 64
Anta = Tapirus terrestris
Anu preto = Crotophaga ani
Apistogramma spp. 70
Aquidauania 60, 64
Arachis glabrata (Amendoim do campo) 53
Aramus guatuana (Carão) 81
Arapaima gigas (Pirarucu) 59, 66
Arapapá = Cochlearius cochlearius
Arara azul = Anodorhynchus hyazinthinus
Ardea cocoi (Garça cinza) 77, 81

Argiopidae 48
Argyrodiaptomus sp, 61
Ariranha = Pteronura brasiliensis
Aristida capillacea (Capim carona) 53
Aristida pallens (Barba de bode) 53
Arroz bravo = Oryza subulata
Asolene platae 64
Aspidosperma cf. ulei (Quebracho blanco) 56
Asplanchna sieboldi 61
Assa peixe = Vernonia sp.
Astronotus ocellata (Cará açu) 70, 75
Attaleia spp. (Acuri) 50
Attheyella sp. 62
Attini 55
Atya paraguayensis 65
Azolla filicoides 48, 51
Bacuparí do rio = Rheedia brasiliensis
Bagre surubim = Pimelodus maculatus
Balsamo = Pterogyne nitens
Barba de bode = Aristida pallens
Barbado = Pinirampis sp.
Barbatimão = Styphnodendrum obovatum
Barriguda = Ceiba pentandra
Bats 88
Beija flores = Hummingbrids
Belostomatidae 63
Biguá = Phalacrocorax olivaceus
Biguá tinga = Anhinga anhinga
Biomphalaria sp. 64
Blastocerus dichotomus (Cervo pantaneiro) 87
Boa constrictor (Jibóia) 76
Bocaiuva = Acrocomia sclerocarpa
Bocuda = Bothrops neuwiedii
Boipevuçu = collective for Mastigodryas and Hydronastes
Bosmina longirostris 61
Bosminopsis sp. 61

*(Portuguese names are also given)

Bothrops neuwiedii (Bocuda) 77
Botryococcus 44
Brachiaria sp. 53
Brachionus sp. 60, 61
Bromelia sp. (Caraguatá) 56, 73
Brucella abortum 85
Brycon orbignianus (= hilari) (Piraputanga) 59, 68, 75
Brysonima intermedia (Canjiquera) 53
Bubulcus ibis (Garça companheira) 83
Buffaloes 88
Bufo paracnemis (Cururu açu) 73
Bugío preto = Alouatta caraya
Burití = Mauritia vinifera
Busarellus nigricollis (Gavião velho) 81
Buteo albicaudatus (Gavião fumaça) 57
Buteogallus urubutinga (Gavião preto) 81
Byssanodonta (= Eupera) paranensis 65
Cabeça seca = Mycteria americana
Cabomba australis 51
Cabomba piauhyensis 51
Cachorro do mato = Dusicyon thous
Cachorro vinagre = Speothos vinaticus
Caesalpinia paraguariensis 56
Caiarana = Guarea microphylla
Caiman 40, 73, 75–76 ff, 96
Caiman latirostris (Jacaré de papo amarelo) 73
Caiman yacare (Jacaré, Jacaretinga) 74, 75–76 ff, 96
Caitetu = Tayassu taiacu
Calango verde = Ameiva ameiva
Cambará = Vochysia rufa
Campephilus robustus (Pica pau rei) 83
Canafistula = Pithecolobium multiflorum
Canjiquera = Brysonima intemedium
Capim carona = Aristida capillacea
Capim colonia = Panicum maximum
Capim d'água = Paspalum repens
Capim mimoso vermelho = Setaria geniculata
Capivara = Hydrochoerus hydrochoeris
Cará açu = Astronotus ocellatus
Caracará = Polyborus plancus
Caraguatá = Bromelia sp.
Caramujo = collective name for crabs
Caramujeiro = Rosthramus sociabilis
Carandá = Copernicia alba
Carão = Aramus giatuana
Cardeal = Paroaria sp.
Cariama cristata (Seriema) 83, 84
Caryocar brasiliense (Piquí) 55

Cascudo = collective name for Plecostomus spp.
Castalia ambigua 64
Catfish 70
Cathartes spp. (Urubu) 83
Catoprion mento 69
Caturitá = Myopsitta monacha
Cavallinesia sp. 56
Cebus apella 88
Ceiba pentandra (Barriguda) 56, 57
Ceratopogonidae 68
Ceratopteris pteridoides 51
Cereus bonplandii (Mandacuru) 56, 57
Ceriodaphnia cornuta 61
Cervo pantaneiro = Blastocerus dichotomus
Chaoboridae 63
Characeae 45
Charadriidae 82, 84
Chauna torquata (Tachã) 81, 82
Cheirodon sp. 69
Chiguire = Spanish for Capybara
Chironomidae 63
Chrysocyon brachyurus (Lobo guará) 87
Chydorus cf. ventricosus 62
Cichlidae, 62, 70
Ciconia maguari (João grande) 78
Circus buffoni 81
Cloudina 9
Coccolobium cujabensis 43, 56
Cochlearius cochlearius (Arapapá) 77
Codorna = Nothura spp.
Colhereiro = Ajaia ajaja
Collosoma macropomum (Tucunaré) 72
Collosoma sp. 69
Combretum leprosum 56
Copaifera sp. (Pau d'oleo) 56
Copernicia alba (Carandá) 50, 53, 55
Corixidae 63
Cornops aquaticus 48
Coruja do campo = Speotyto cunicularis
Corumbatá = Prochilodes reticulatus
Cosmedorius albus (Garça alba) 77, 81
Croata = Dyckia sp.
Crotophaga ani (Anu preto) 83
Cuica = collective name for Didelphidae
Cuniculus paca (Paca) 87
Curatella americana (Lixeira) 55
Curculionidae 48
Curicacá = collective name for ibises
Curicacá cinza = Theristicus caerulescens
Curimata spilura 69
Cururu açu = Bufo paracnemis

Cutía = Dasyprocta variegata
Cyanobacteria 44
Cyclestheria hislopi 60, 63
Cyperus giganteus 52
Cyperus haspan 49
Cyprinodontidae 68
Daphnia spp. 61
Dasyprocta variegata (Cutía) 87
Dendrocygna bicolor (Marreca caneleira) 81
Derallus rudis 62
Dermatonotus sp. 73
Dero spp. 62
Desmidiaceae 44
Diaphanosoma spp. 61, 62
Diaptomidae 61, 62, 63
Didelphis sp. (Gambá) 88
Dilocarcinus pagei 65
Diplodon paranense 64
Diptychandra glabra 55
Dolichonyx oryzivorus 84
Dourado = Salminus maxillosus
Dracaena paraguayensis (Vibora do Pantanal) 59, 78
Drepanotrema sp. 64
Dryopidae 63
Duhnevendia odontoplax 62
Dusicyon thous (Cachorro do mato) 76, 87
Dyckia (Croata) 56
Dytiscidae 634
Echinodorus spp. 51
Ectocyclops phaleratus 62
Ectocyclops rubescens 62
Eichhornia (Aguapé) 38, 70
Eichhornia azurea 52
Eichhornia crassipes 46, 47, 48, 49, 50, 52, 62
Elocharis sp. 49
Elodea (= Anacharis) ernstae 51
Ema = Rhea americana
Enterolobium contortisiliquum (Orelha de pau) 56
Ephemeroptera 63
Erythroxyllum tuberosum (Mercurio bravo) 53
Espichadeira = Solanum malacoxylon
Eucyclops neumani 62
Euglenacea 44
Euglena sanguinea 44
Eunectes notaeus (Sucurí) 76
Euryalona fasciculata 62
Eurypyga helias (Pavãozinho do Pará) 82
Felis concolor (Onça parda) 88
Felis pardalis (Jaguatirica) 88

Felis tigrina (Gato do mato pequeno) 88
Felis wiedii (Gato maracajá) 88
Felis yagouaroundi (Gato mourisco)
Feral animals 82, 93
Ficus elliotiana 54
Ficus spp. (Mata pau) 55
Fluvicola leucocephala (Maria velhinha) 84
Fluvicola pica (Lavandeira) 82, 84
Fossula fosculifera 64
Frango d'água azul = Porphyrula martinica
Fruta do lobo = Solanum sp.
Fumo bravo = Polygonum hispidum
Gallinula chloropus (Frango d'água) 81
Gambá = Didelphis sp.
Garça alba = Cosmedorius albus
Garça cinza = Ardea cocoi
Garça companheira = Bubulcus ibis
Garça real = Ptherodius pileatus
Gastropelecus sternicla 69
Gato do mato pequeno = Felis tigrina
Gato mourisco = Felis yagouaroundi
Gavião fumaça = Buteo albicaudatus
Gavião preto = Buteogallus urubutinga
Gavião velho = Busarellus nigricollis
Geochelone sp. 77
Geophilus spp. 70
Gir cattle 94
Grundulus sp. 59
Guarea microphylla (Caiarana) 54
Guaximim = Procyon cancrivorus
Gundlachia 64
Gymnotidae (Tuvira) 68
Harbenia aricaensis 46, 47, 52
Harpacticoida 61, 62
Helicops leopardina 77
Helminths 86
Heliornis fulica (Ipequi) 82
Hemiptera 63, 75
Hesperidae 58
Heteranthera limnosa 52
Hirudinea 63
Hoplias malabaricus (Traíra) 68, 71, 75
Horaella thomassoni 60
Horses 88, 89
Hummingbirds 83
Hydrochoerus hydrochaeris (Capivara) 85–86 ff
Hydrocleys spp. 51
Hydrodynastes gigas (Boipevuçu) 77
Hydrolea spinosa 51
Hydrophilidae 63
Hydroptylidae 63
Hydrovatus sp. 63

Hyla, Hylidae (Perereca) 73
Hymenaea courbarii (Jatobá) 56
Hymenaea stigonocarpa (Jatobá) 55
Hyphessobrycon flameus (Lambari) 66
Hypherssobrycon serpae (Lambari) 69
Hypoclinemus sp. 59, 67
Hyriidae 64, 65
Ibises (Curicacá) 78
Ierba mate 93
Iguana spp. 76
Iheringiella balzani 64
Ilyocryptus sordidus 62
Ilyocryptus spinifera 61
Indialona globosa 62
Inga spp. 54
Ipê amarelo = Tabebuia aurea
Ipequí = Heliornis fulica
Ipomoea sp. 48
Ipomoea fistulosa (Algodão bravo) 53
Jabiru mycteria (Tuiuiu) 78
Jacana jacana 48, 82
Jacaretinga = Caiman yacare
Jacaré coroa = Paleosuchus palpebrosus
Jacaré de papo amarelo = Caiman latirostris
Jaguatirica = Felis pardalis
Jaracaca = Bothrops neuwiedii
Jatobá = Hymenaea courbarii, H. stigonocarpa
Jaú = Paulicea lutkeni
Jiboia = Boa constrictor
João grande = Ciconia maguari
Juazeiro = Zyziphus joazeiro
Jurubeba = Solanum sp.
Killmeyera coriacea (Pau santo) 55
Kingfishers 82
Lamproscapha ensiformis 65
Lavandeira = Fluvicola picta
Leaf cutter ants 55
Lebias sp. 69
Lecane (Monostyla) copeis 60
Lecane spp. 61
Leila blainvilleana 64
Lemna sp. 52
Lepidosiren paradoxus (Pirambóia) 38, 72
Leptocinclis 44
Leporinus friderici (Piavuçu) 65, 72
Leptodacylidae 73
Leptodactylus sp. (Rã pimenta) 73
Leydigiopsis curvirostris 62
Leydigiopsis ornata 62
Licania sp. 54
Limnobium stoloniferum 51
Liophis miliaris 77

Littoridina spp. 64
Lixeira = Curatella americana
Lobo guará = Crysocyon brachyurus
Lontra = Lutra platensis
Loricariidae 70
Ludwigia (= Jussiea) spp. 49, 51, 62
Lungfish 67, 72
Lutra platensis (Lontra) 85
Lysapsus limellus 73
Macrobrachium amazonicum 65
Macrocyclops albidus 62
Macrothricidae 61, 62
Mandacuru = Ceriops bonplandii
Marelia remipes 48
Maria velhinha = Fluvicola pica
Marisa planogyra 64
Markiana sp. 59
Marreca caneleira = Dendrocygna bicolor
Marreca de coleira = Anas leucophrys
Marsilea polycarpa 51
Mastigodryas bifossatus (Boipevuçu) 77
Mata pau = Ficus sp.
Mauritia vinifera (Buriti) 50
Mazama americana (Veado catinguiero) 87
Mentzelia corumbaensis 49, 56
Mercurio bravo = Erythroxyllum tuberosum
Mergulhão pompom = Podiceps dominicus
Mesocyclops longisetosus 62
Mesocyclops meridionalis 62
Metacyclops mendocinus 61
Metacyclops sp. 62
Micractinium 44
Microcyclops anceps 62
Microcyclops ceibaensis 62
Microcyclops finitimus 62
Microcyclops sp. 61
Microcyclops varicans 62
Microcystis aeruginosa 44
Microhylidae 73
Milvago chimachima (Pinhé) 83
Mimosa sp. 56
Moenkhausia spp. 69
Moina minuta 61
Morcego pescador = Noctilio leporinus
Mussum = Symbranchus vulgaris
Mutelinae, Mutelaceae 64, 65
Mycteria americana (Cabeça seca) 78, 81
Myloplus 69
Mylosoma 69
Myopsitta monacha (Caturitá) 78, 83
Myrmecophaga tridactyla (Tamanduá bandeira) 88
Nasua nasua (Quatí) 76, 81, 88

Nectomys (Rato d'água) 87
Nelore cattle 93–94 ff
Nematoda 63
Neochen jubata (Pato corredor) 81
Neochetina eichhorniae 48
Nocitilio leporinus (Morcego pescador) 88
Noteridae 63
Nothura sp. (Codorna) 83
Notodiaptomus coniferoides 61
Notodiaptomus spp. 61
Nycticorax nycticorax (Socó dorminhoco) 77, 81
Nymphaea amazonum 48
Nymphalidae 58
Nymphoides spp. 52
Odonata 63
Oedogonium 44
Oligochaeta 64
Onça parda = Felis concolor
Onça pintada = Panthera onca
Onze horas = Portulaca sp.
Oocystis 44
Orelha de onça = Salvinia
Orelha de pau = Enterolobium contortisiliquum
Orthoptera 48
Oryza perennis 89
Oryza subulata (Arroz bravo) 53
Oscillatoria limnetica 44
Ostracoda 62
Oxyethira sp. 63
Oxyurella longicaudis 62
Ozotocerus bezoarticus (Veado campeiro) 87
Paca = Cuniculus paca
Pacyhurus 67
Pacu = Piaractus mesopotamicus
Palaemonetes spp. 65
Paleosuchus palpebrosus (Jacaré coroa) 73
Palometa = Spanish for Piranha
Pandion haliaetus (Águia pesqueira) 81
Panicum maximum (Capim colonia) 57
Panthera onca (Onça pintada) 87
Papilionidea 58
Paracyclops fimbriatus 62
Paratheria prostata (Capim mimoso) 52
Paratrygon spp. 58, 59
Paratudo = Tabebuia caraiba
Paroaria capitata (Cardeal) 84
Parodon sp. 68
Paspalum repens (Capim d'água) 49, 52
Pato corredor = Neochen jubata
Pau de formiga = Triplaria formicosa
Pau d'oleo = Copaifera sp.

Pau santo = Kielmeyera coriacea
Pau terra = Qualea parviflora
Pau de tucano = Vochysia tucanorum
Paulicea lutkeni (Jaú) 66, 67, 70, 72
Paulinia acuminata 48
Pavãozinho do Pará = Eurypyga helias
Pediastrum duplex 44
Perereca = collective name for Hylidae
Pereskia saccharosa 56
Phaetornis spp. 83
Phalacrocorax olivaceus (Biguá) 77, 81
Phimosus infuscatus 78
Phoxinopsis sp. 59
Phrynops geoffroanus 77
Phyllanthus fluitans 51
Phyllomedusa hypochondrialis 73
Physolaemus biliginogerus (Rã de brejo) 73
Piaractus mesopotamicus (Pacu) 67, 69, 72
Piavaçu = Leporinus friderici
Pica pau rei = Campephilus robustus
Pimelodes maculatus (Bagre surubim) 70
Pinhé = Milvago chimichima
Pinirampus sp. (Barbado) 72
Pintado = Pseudoplatysoma coruscans
Piptadenia macrocarpa (Angicō) 56
Piquí = Caryocar brasiliense
Piramboia = Lepidosiren paradoxus
Piranha = collective name for Serrasalminae, 59, 70 ff, 72, 81
Piraputanga = Brycon orbignianus
Pir = Scirpus validus
Pistia stratiotes (Alface d'água) 46, 48, 49, 50, 51
Pithecolobium multiforum (Canafistula) 54
Pitu = collective name for Palaemonidae
Plagoscion spp. 67
Planorbidae 64
Platemys macrocephala 77
Platyas leloupi 60
Plecostomus plecostomus (Cascudo) 70, 75
Pleidae 63
Podiceps dominicus (Mergulhão pompom) 77
Polyborus plancus (Caracará) 83, 84
Polygonum hispidum (Fumo bravo) 52
Pomacea escalans 64
Pomella sp. 64
Pontederia lanceolata 52
Poppiana argeninianus 65
Porco monteiro = Feral pig 87
Porphyrula martinica (Frango d'água azul) 82
Portulaca spp. 53

Prochilodes reticulatus (Corumbatá) 67, 68, 72
Procyon cancrivorus (Guaximím) 88
Prosopis sp. 56
Pseudis paradoxus 73
Pseudopaludicola ameghini 73
Pseudoplatystoma corruscans (Pintado) 67, 70, 72
Pseudosida sp. 61
Pterocarpus rohri 54
Pterogyne nitens (Balsamo) 55
Pteronura brasiliensis (Ariranha) 85
Ptherodius pileatus (Garça real) 77
Pygocentrus (= Serrasalmus) nattereri (Piranha) 69, 70
Pygopristis sp. 70.
Pyrrhulina brevis australe, 69, 71
Qualea grandiflora 55
Qualea parviflora (Pau terra) 55
Quatí = Nasua nasua
Quebracho blanco = Aspidosperma ulei
Quebracho colorado = Schinopsis sp.
Queixada = Tayassu albirostris
Rã de brejo = collective name for Physalaemus
Rã pimenta = collective name for Leptodactylus
Rhamphastos toco (Tucanuçu) 83
Rails (Saracurá) 81
Rato d'água = Nectomys sp.
Reussia rotundifolia 48, 52
Reussia subovata 48, 52
Rhamdia pubescens 70, 75
Rhea americana (Ema) 83
Rheedia brasiliensis (Bacuparí do rio) 54
Rhynchospora corymbosa 49
Rissoidea 64
Rivulinae 68
Roestes sp. 68
Rosthramus sociabilis (Caramujeiro) 81
Rotifera, Rotatoria 60, 61, 63
Rotunditermes bragantinus 55
Salminus maxillosus (Dourado) 68, 69
Salvinia auriculata (Orelha de onça) 47, 48, 50, 51, 62, 63
Sapo = Bufo paracnemis
Saracuras = Rallidae, Rails
Sarkidiornis melanotos 81
Scenedesmus 44
Schinopsis sp. (Quebracho colorado) 56
Schleschiella 65
Sciaenidae 67
Scirpus cubensis 49

Scirpus validus (Pirí) 52
Senna pendula 49
Seriema = Cariama cristata
Serrasalminae (Piranhas) 59, 69, 73
Serrasalmus marginatus 70
Serrasalmus spilopleura 70, 75
Setaria geniculata (Capim mimoso vermelho) 53
Siluriformes 67
Simocephalus sp. 62
Siphonops sp. 73
Socó boi = Tigrisoma lineatum
Socó dorminhoco = Nycticorax nycticorax
Speleogryphaceae 39
Speothos vinaticus (Cachorro vinagre) 87
Speotyto cunicularis (Coruja do campo) 83
Sphaeriidae 65
Spirogyra 44
Spirulina 44
Stingrays (Paratrygon) 67
Stratiomyidae 63
Strombomonas 44
Strongylura 67
Stryphnodendrum obovatum (Barbatimão) 55
Sucurí = Eunectes notaeus
Suphisellus grammicus 62
Sylviocarcinus spp. 65
Symbranchus vulgaris (Mussum) 67
Tabebuia aurea (Ipê amarelo) 54
Tabebuia caraiba (Paratudo) 50, 54, 55
Taboa = Typha dominguensis
Tachã = Chauna torquata
Tachycineta albiventer (Andorinha do rio) 82
Tamanduá bandeira = Myrmecophaga tridactyla
Tapirus terrestris (Anta) 87
Tarumá = Vitex cymosa
Tayassu albirostris (Queixada) 87
Tayassu tajacu (Caitetu) 87
Teju = Tupinambis teguixim
Terapins 40
Testudinella ahlstromi 60
Tetras 69
Thalia geniculata 52
Theristicus caerulescens (Curicacá cinza) 78
Thermocyclops minutus 61
Tigrisoma lineastum (Socó boi) 78
Traíra = Hoplias malabaricus
Trichechus triunguis 59, 85
Trichocerca chattoni 60
Trichodactylidae 65, 75
Trichodactylus spp. 65

Trichoptera 63
Triplaris formicosa (Pau de formiga) 54
Tropisternus collaris 62
Trypanosoma evansi 85
Tucunaçu = Rhamphastos toco
Tucunaré = Collosoma macropomum
Tucura cattle 93
Tuiuiu = Jabiru mycteria
Tupinambis teguixim (Tejú) 76
Turbellaria 63
Tuvira = Gymnotus
Typha dominguensis (Taboa) 52
Urubu = Cathartes spp.
Utricularia spp. 46, 52
Veado campeiro = Ozotocerus bezoarticus

Veado catingueiro = Nazama simplicicornis
Veado mateiro = Mazama americana
Vernonia spp. (Assa peixe) 53
Vibora do Pantanal = Dracaena paraguayensis
Victoria cruziana 46, 52
Vitex cymosa (Tarumá) 55
Vochysia rufa (Cambará) 55
Vochysia spp. 54
Vochysia tucanorum (Pau de tucano) 55
Xenurobrycon spp. 59
Xiliphus sp. 59
Zilchopsis sattleri 65
Zyziphus joazeiro 56

Geographical and Ethnographic Index*

Acre 9
Acurizal 43, 56, 84
Amazon 4, 9, 20, 22, 38, 41, 59, 61, 64, 65, 66, 67, 69, 70, 73, 85, 101
Andes(andine) 8, 9, 10
Arawak 89, 91
"Archiplata" 10, 60
Argentine 3, 47, 49, 50, 75
Asuncion 4, 89
Bahia Negra (Paraguay) 31, 76, 101
Báa do Arame 44, 45, 61
Baía do Jacaré 44, 61
Baía Grande 22 44
Bananal (Island of) 37
    Banhados do Izozog 37
Barão do Melgaço 29
Bolivia 1, 3, 6, 34, 37, 41.53, 55, 75, 76, 92, 97, 101
Bonito 10
Bororós 89, 92, 95
Brazilian Highlands (Planalto) 10, 17, 41, 43, 55, 99, 100
Cáceres 12, 20, 29, 92, 99
Cadiveus 90, 91
Campo Grande 3, 17, 97, 101
Canal Dom Pedro II 29
Caracará National Park 101
Cassiquiare canal 22
Chaco 29, 32, 34, 39, 41, 56, 57, 58, 73, 76, 81, 85
Chamacocos 90
Chapada dos Guimarães 6, 37
Chapada dos Parecis 6, 19
Coimbra 101
Coricha Grande 8, 29, 34

Corumbá 3, 37, 92, 96, 97, 98, 101, 102
Coxim 6
Cuiabá 4, 5, 24, 37, 63, 64, 89, 92, 96, 98, 100, 101
Eupana Lacus 3
Fazenda Bodoquena 92
Fazenda Caiamá 102
Fazenda Firme 92
Fazenda Nhumirim 28, 61, 71
Fecho dos Morros 6, 12, 20, 25
Fonte Progresso 10
Gondwana 8
"Great Southamerican Disjunction" 43
Guaicurus 4, 89, 90, 91, 93
Guanás 89, 90, 91
Guatós 89
Guayaquil gap 9
Humaitá 37
Ladário 24, 27
Lagoa Agostinho 37
Lagoa da Rebeca 22
Lagoa Negra 29
Lagoa, see also Lake
Lake Cáceres 29
Lake Chocororé 29
Lake Gaiba 4, 20, 29, 43, 101
Lake Jacadigo 29, 60, 61
Lake Mandioré 29
Lake Orion 8, 29
Lake Regagua 37
Lake Regaguado 37
Lake Uberaba 4, 8, 20, 29
La Plata 1, 20, 22, 38, 40, 48, 49, 59, 60, 93
Llanos de Moxos 37
Mbayás 89, 90

*(Very frequent items like Pantanal, Rio Paraguay, Mato Grosso, etc. are not included)

Melgaço 5
Mimoso 91
Minas Gerais 95
Miranda 76, 79, 92
Mkhtlawaiya (Paraguay) 37
Morro do Azeite 16, 56
Nabileque (region) 34, 35
Nambikwaras 91
Nhecolândia 14, 27, 28, 30, 31, 32, 43, 45, 60, 63, 70, 92, 101
Paiaguás (Indian nation) 4, 89
Paiaguás (region) 28, 35
Paraguay 1, 3, 6, 7, 19, 25, 41, 49, 67, 75, 76, 101
Paraná basin 8, 20, 24
Pebas formation 9
Poconé 32, 35, 44, 60, 61, 62, 76, 79, 92, 99
Porto Jofre 31, 32, 37, 63, 98
Porto Murtinho 19, 24
Rio
   Aguapei 22
   Alegre 22
   Apa 6, 17, 43
   Aquidauana 7, 12, 20, 21, 27, 34
   Araguaia 27, 37
   Beni 75
   Bento Gomes 21, 99, 100
   Cabeçal 19
   Caracará 21
   Casange 21
   Coxim 99
   Cuiabá 7, 12, 20, 21, 24, 25, 27, 29, 31, 32, 34, 37, 72, 99
   das Mortes 37
   das Petas 8
   Formoso 10
   Guaporé 6, 22, 37, 75
   Itiquira 21, 102
   Jauru 7, 12, 19, 20, 22, 68, 73
   Madeira 3, 99
   Magdalena 59
   Mamoré 37, 60, 75

Manso 99
Miranda 7, 10, 12, 13, 15, 16, 20, 21, 25, 27, 31, 32, 34, 43, 56, 63
Mogi Guaçu 67
Nabileque 17, 20, 43
Negro 21, 22, 92
Negro (Amazonia) 22, 65
Orinoco 22, 59, 61
Otoquis 8, 37, 75
Paraguai Mirim 20
Paraná (river and basin) 32, 33, 38, 48, 49, 61, 63, 67, 73
Perdido 22
Piquiri 37, 102
Pilcomayo 67
Quimome 37
São Francisco
São Lourenço 7, 12, 20, 21, 29, 34, 35
Sepetuba 19
Taquari 7, 11, 12, 20, 21, 25, 28, 34, 72, 92, 99
Tocantins 27, 37, 66
Rio de Janeiro 5
Salina do Meio (Nhumirim) 44
Santa Cruz (Bolivia) 37, 75
São Paulo state 4, 92, 95
Sea of Xaraes 4, 12
Serra da Bodoquena 6, 31, 39, 41, 90
   do Amolar 101
   do Chocororé
   do Maracaju 6
   do São Jeronimo 6
   do Urucum 9, 97
Sudd 1, 50
Taiamá (Island of) 101
Transpantaneira road 32, 37, 97, 98
Terenos 90
Tupi(-Guaraní and Kawahib) 89, 91
Uruguay 3
Xaraes (tribe) 4
Xaraes Formation 11

Subject Index*

Agriculture 99
Alkaline waters 31
Allochthonous river 19
Amazonian forest 41–43, 56
Anthropic changes 3, 99
Aquatic plants 38, 40, 43, 45–50
Aquifers 20–23
Arid, aridity 1, 7, 12, 16–19, 28, 34, 38, 39
Aterros 99
Atlantic forest 41, 43, 58, 59
Baías 15, 22, 28–31, 60, 61, 70, 71, 75
Baixada 16
Bandeirantes 4, 92
Banhado, 16
Barreiros, 16, 92
Barrier, biogeographic 39, 58
Batume 29, 48, 49
Biogeography 39, 41, 58–60
Blackwaters 31, 37, 66
Boca 14
Boiadeiros 95
Brackish waters 8, 9, 71
Brasil, Vital 5
Cabeza de Vaca, Nunes 4
Calcareous rocks, 10, 29, 31, 39, 42, 56
Camalotes 49–50, 62, 63, 70, 75
Campos 42, 43, 52–53, 55, 83
Capoes 16
Cattle, cattle farming 24, 27, 32, 53, 85, 93–95
Cerradão 49, 55, 57
Cerrado 41–43, 55, 57, 58, 84, 85, 87, 88
Chaco biota 39, 41–43, 55, 56, 58, 73, 81, 85
Chlorophyll 32
Climate 1, 6, 7, 12, 16–19, 24, 30, 31, 34, 39
Coivaras 39
Collector rivers 6, 8, 12, 19–22, 25
Colonists, colonization 4, 89–93

Cones, alluvional 11, 12, 13, 25
Conductivity 31, 32
Conservationism 3, 53, 98–103
Cordilheiras 16, 30, 85
Corixo 14, 45, 48
Corrego 14
Coureiros 76, 96
Cretaceous period 8, 10
Cruz, Oswaldo 5
Cultivation 89, 92, 95
Cycles, climatic 23–28, 29, 32, 44, 71
Dequada 67
Diques 16 Diversity 39, 43, 44, 45, 59, 63, 64, 70, 72, 87, 98, 102
Drillings 10, 11
Droughts 12, 27, 28, 38, 49, 54
Dry season, drying-out 12, 15, 22, 28, 29, 32, 38, 44, 48, 61, 75, 94
Ecotourism 97, 101, 102
El Nino (ENSO) 27
Endemism 39, 40, 43, 44, 45, 54, 56, 58–60, 64, 67, 76, 77, 78, 81, 83
Endorheic basins 7, 11, 19, 29, 100, 112
Eocene 9
Escarpments 6, 9, 12
Evaporation, evapotranspiration 17, 19, 21
Expeditions 4, 5
Fans, alluvional 12, 42
Ferreira, Alexandre Rodrigues 4
Fire 28, 38, 55, 57, 84, 89
Fishermen, fishing 19, 24, 39, 65, 69, 72, 95, 96
Floating plants 29, 38, 40, 45–50, 60–63, 65, 68–70, 73, 75, 82, 98
Floods 1, 6, 8, 16, 19, 20, 23–28, 34, 37, 49, 61, 94, 99
Forecasting floods 23–27
Fossils 9, 11

*(Local terminology is also included as well as names of explorers)

Friagens 17, 58
Gallery forest (Mata ciliar) 4, 12, 16, 39, 41, 54, 55, 79, 83, 85, 88
Garimpeiros 96, 99
Glacialis 12, 42
Godichaud, Charles 5
Gold mining 4, 12, 96, 99
Graben 9, 10
Grasslands 1, 3, 7, 42, 52, 57, 100
Headwaters 10, 14, 19, 20, 22, 24, 25, 31, 37, 66, 72
Herbaceous vegetation 52, 55
Herbicides 100
"Hidrovia Project" 3, 99
Hoehne, F. 5
Humidity, humid phases 1, 12, 22
Hunting, hunters 73, 75, 76, 85
Hydatophytes 44, 46, 47
Hydroelectric plants 99
Indians, indian tribes 4, 89–91
Iron ore 9, 97
Lagoon, brackish 8
Lambedouro 15
Langsdorff, Georg Heinrich von 4, 5
Largo 16
Lufada 66, 67
Manganese ore 9, 97
Marine, marine faunal elements 9, 65–67
Mata ciliar (see also gallery forest) 54
Meiofauna 60–64
Mercury 99, 100
Migrations 38, 66–68, 70, 81–83, 84, 99
Monadnocks 1, 9, 56
Moore, Nerchant 5
Morro's, morraria 15, 45, 69, 70
Murundu's 54, 55
Natterer, Joseph 5 Navigation 3, 20, 92, 94
Neogene 8
Neotropics, neotropical 41, 61, 73, 77, 82
Ninhals (rookeries) 79–81, 83, 87, 102
Nitrogen, total 32
Nutrients 32, 44, 61
Orbigny, Alcide d' 5
Overflow lakes 12, 24, 29
Oxbow lakes 29, 37
Oxygen 31, 38, 45, 46, 50, 62, 67, 69, 70
Palaeozoic 9
Pantanal, limits 6–8
Pantanal, size 6
Pantaneiros 14, 57, 88, 89, 95, 102
Paraguayan War 4, 92, 92
Parques, "parks" 1, 50

pH 31, 100
Phosphate 39
Phytal fauna 61, 62–63
Phytogeography 12, 41–44
Phytoplankton 44
Piracema 38, 66, 67, 72, 99
Pleistocene 11, 13, 39, 42, 57
Pleuston 48, 62
Pliocene 12, 14, 59
Pluvial lakes 29
Poaching 75, 76, 85, 96, 102
Pollution 99, 100
Postglacial 12, 42
Precambrian 16, 30, 56
Precipitations 57
Primary production 44, 48
Pyrophytes 57
Quebracho shrub 56, 57
RADAMBrasil programme 1
Rainforest 39, 43, 54, 58, 73
Refugia 12, 55
Ridges, mountain 6, 7, 9, 16, 20
Roads 11, 37, 72, 83, 97, 98
Rondon, Cândido 5, 91
Roosevelt, Theodore 5, 67
Round baías (lakes) 31, 32, 44, 61, 63, 68, 70, 76
Salinas 31, 32, 44, 61, 63, 68, 70, 76
Salinity 29, 37, 71, 76
Salt pans 1, 7, 32
Sangradouros 14
Savanna 16, 37, 39, 41, 42, 55, 83, 95
Seasons, hydrological 8, 11, 16, 20, 22, 25, 29, 31, 32, 37, 39, 44, 52, 55, 57, 61, 75, 79, 85, 101
Sediments 9–12, 22, 31, 56, 99
Semi-deciduous forests 39, 41, 55, 56
Serra, serrania (see also geographic index) 16
Sesmarias 92, 95
Shrubland 39, 48, 55, 57
Sod lakes 31, 32, 44 Soils 30, 31, 34
Steinen, Carl von 5
Subsidence 9, 11
Succession, biological 39, 42, 43, 48, 50, 55
Sumidouros 22
Syncline 1, 10
Tectonism 9
Temperatures 12, 16, 17, 37, 38
Temporary streams, pools 6, 7, 12, 37, 68
Torrents 6, 11, 14, 21, 22, 39
Travertine 6, 11, 14, 21, 22, 39

Várzea 16, 32, 61
Vertentes 14
Vicariance 59
Wetlands 1, 5, 19, 31, 37, 38

Xerophytes 56
Zoogeography 40, 58–60
Zooplankton 60–62

MONOGRAPHIAE BIOLOGICAE

1. *Physiologia Comparata et Oecologia*, Vol. I. 1948   out of print
2. *Physiologia Comparata et Oecologia*, Vol. II. 1952   out of print
3. *Physiologia Comparata et Oecologia*, Vol. III. 1954   out of print
4. *Physiologia Comparata et Oecologia*, Vol. IV. 1957   out of print
5. (1) R. A. Kennedy: *The Metzner Theory of Urine Formation.* 1957
   (2) M. D. Grmek: *On Ageing and Old Age.* Basic Problems and Historic Aspects of Gerontology and Geriatrics. 1958
   (3–4) O. Bassir: *Biochemical Aspects of Human Malnutrition in the Tropics.* 1962
       out of print
6. F. S. Bodenheimer: *Animal Ecology Today.* 1958   out of print
7. R. R. Gates: *Taxonomy and Genetics of Oenothera.* 1958   ISBN 90-6193-061-8
8. A. Keast, R. L. Crocker and C. S. Christian (eds.): *Biogeography and Ecology in Australia.* 1959   out of print
9. S. Stanković: *The Balkan Lake Ohrid and Its Living World.* 1960   out of print
10. E. Rivnay: *Field Crop Pests in the Near East.* 1962   ISBN 90-6193-063-4
11. M. Kozhov: *Lake Baikal and Its Life.* 1962   ISBN 90-6193-064-2
12. M. S. Mani: *The Ecology of Plant Galls.* 1964   out of print
13. E. van Oye (ed.): *The World Problem of Salmonellosis.* 1964
        ISBN 90-6193-066-9
14. D. H. S. Davis (ed.): *Ecological Studies in Southern Africa.* 1964   out of print
15. J. van Mieghem and P. van Oye (eds.): *Biogeography and Ecology in Antarctica.* 1965   ISBN 90-6193-067-7
16. H. Boyko (ed.): *Salinity and Aridity.* New Approaches to Old Problems. 1966
        out of print
17. M. J. Coe: *The Ecology of the Alpine Zone of Mount Kenya.* 1967
        ISBN 90-6193-068-5
18. E. J. Fittkau, J. Illies, H. Klinge, G. H. Schwabe and H. Sioli (eds.): *Biogeography and Ecology in South America*, Vol. I. 1968   out of print
19. E. J. Fittkau, J. Illies, H. Klinge, G. H. Schwabe and H. Stioli (eds.): *Biogeography and Ecology in South America*, Vol. II. 1969   ISBN 90-6193-071-5
20. R. N. Kaul (ed.): *Afforestation in Arid Zones.* 1970   out of print
21. R. Battistini and G. Richard-Vindard (eds.): *Biogeogaphy and Ecology of Madagascar.* 1972   ISBN 90-6193-073-1
22. J. E. Bishop. *Limnology of a Small Malayan River Sungai Gombak.* 1973
        ISBN 90-6193-074-X
23. M. S. Mani (ed.): *Ecology and Biogeography in India.* 1974   ISBN 90-6193-075-8
24. E. K. Balon and A. G. Coche (eds.): *Lake Kariba.* A Man-made Tropical Ecosystem in Central Africa. 1974   ISBN 90-6193-076-6
25. W. D. Williams (ed.): *Biogeography and Ecology in Tasmania.* 1974
        ISBN 90-6193-077-4
26. A. de Vos: *Africa, the Devastated Continent?* Man's Impact on the Ecology of Africa. 1975   ISBN 90-6193-078-2
27. G. Kuschel (ed.): *Biogeography and Ecology in New Zealand.* 1975
        ISBN 90-6193-079-0
28. I. Prakash and P. K. Ghosh (eds.): *Rodents in Desert Environments.* 1975
        ISBN 90-6193-080-4
29. J. Rzóska (ed.): *The Nile.* Biology of an Ancient River. 1976   ISBN 90-6193-081-2
30. G. Kunkel (ed.): *Biogeography and Ecology in the Canary Islands.* 1976
        ISBN 90-6193-082-0

MONOGRAPHIAE BIOLOGICAE

31. M. J. A. Werger (ed.): *Biogeography and Ecology of Southern Africa*, 2 vols. 1978
    ISBN 90-6193-083-9
32. C. Serruya (ed.): *Lake Kinneret* [Lake of Tiberias / Sea of Galilee]. 1978
    ISBN 90-6193-085-5
33. Ph. D. Mordukhaí-Boltovskoi (ed.): *The River Volga and Its Life.* 1979
    ISBN 90-6193-084-7
34. Ch. W. Heckman: *Rice Field Ecology in Northeastern Thailand.* The Effect of Wet and Dry Seasons on a Cultivated Aquatic Ecosystem. 1979   ISBN 90-6193-086-3
35. M. Kalk, A. J. McLachlan and C. Howard-Williams (eds.): *Lake Chilwa.* Studies of Change in a Tropical Ecosystem. 1979   ISBN 90-6193-087-1
36. B. R. Allanson (ed.): *Lake Sibaya* [South Africa]. 1979   ISBN 90-6193-088-X
37. H. Löffler (ed.): *Neusiedlersee.* The Limnology of a Shallow Lake in Central Europe. 1979   ISBN 90-6193-089-8
38. J. Rzóska: *Euphrates and Tigris.* Mesopotamian Ecology and Destiny. With Contributions by J. F. Talling and K. E. Banister. 1980   ISBN 90-6193-090-1
39. K. A. Brodsky: *Mountain Torrent of the Tien Shan.* A Faunistic-Ecology Essay. Translated from Russian by V. V. Golosov. 1980   ISBN 90-6193-091-X
40. M. S. Mani and L. E. Giddings (eds.): *Ecology of Highlands.* 1980
    ISBN 90-6193-093-6
41. A. Keast (ed.): *Ecological Biogeography of Australia*, 3 vols. 1981
    ISBN 90-6193-092-8
42. J. L. Gressit (ed.): *Biogeography and Ecology of New Guinea*, 2 vols. 1982
    ISBN 90-6193-094-4
43. H. Heathwole, T. Done and E. Cameron: *Community Ecology of a Coral Cay.* A Study of One-Tree Island, Great Barrier Reef, Australia. 1981   ISBN 90-6193-096-0
44. P. S. Maitland (ed.): *The Ecology of Scotland's Largest Lochs.* Lomond, Awe, Ness, Morar and Shiel. 1981   ISBN 90-6193-097-9
45. K. Müller (ed.): *Coastal Research in the Gulf of Bothnia.* 1982   ISBN 90-6193-098-7
46. G. K. Rutherford (ed.): *The Physical Environment of the Faeroe Islands.* 1982
    ISBN 90-6193-099-5
47. J. I. Furtado and S. Mori (eds.): *Tasek Bera* [Malaysia]. The Ecology of a Feshwater Swamp. 1982   ISBN 90-6193-100-2
48. G. R. South (ed.): *Biogeography and Ecology of the Island of Newfoundland.* 1983
    ISBN 90-6193-101-0
49. J. A. Thornton (ed.): *Lake McIlwaine.* The Eutrophication and Recovery of a Tropical African Man-made Lake. 1982   ISBN 90-6193-102-9
50. R. W. Edwards and M. P. Brooker: *The Ecology of the Wye* [U.K.]. 1982
    ISBN 90-6193-103-7
51. T. Petr (ed.): *The Purari* [Papua New Guinea]. Tropical Environment of a High Rainfall River Basin. 1983   ISBN 90-6193-104-5
52. H. Kuhbier, J. A. Alcover and C. Guerau d'Arellano Tur (eds.): *Biogeography and Ecology of the Pityusic Islands.* 1984   ISBN 90-6193-105-3
53. J.-P. Carmouze, J.-R. Durand and C. Lévêque (eds.): *Lake Chad.* Ecology and Productivity of a Shallow Tropical Ecosystem. 1983   ISBN 90-6193-106-1
54. S. Horie (ed.): *Lake Biwa* [Japan]. 1984   ISBN 90-6193-095-2
55. D. R. Stoddart (ed.): *Biogeography and Ecology of the Seychelles Islands.* 1984
    ISBN 90-6193-107-X
56. H. Sioli (ed.): *The Amazon.* Limnology and Landscape Ecology of a Mighty Tropical River and Its Basin. 1984   ISBN 90-6193-108-8

MONOGRAPHIAE BIOLOGICAE

57. C. H. Fernando (ed.): *Ecology and Biogeography in Sri Lanka.* 1984
    ISBN 90-6193-109-6
58. S. J. Casper (ed.): *Lake Stechlin* [Germany]. A Temperate Oligotrophic Lake. 1985
    ISBN 90-6193-512-1
59. U. Th. Hammer: *Saline Lake Ecosystems of the World.* 1985   ISBN 90-6193-535-0
60. B. R. Davies and K. F. Walker (eds.): *The Ecology of River Systems.* 1986 out of print
61. P. de Deckker and W. D. Williams (eds.): *Limnology in Australia.* 1986
    ISBN 90-6193-578-4
62. Y. Yom-Tov and E. Tchernov (eds.): *The Zoogeography of Israel.* The Distribution and Abundance at a Zoogeographical Crossroad. 1988   ISBN 90-6193-650-0
63. F. D. Por: *The Legacy of Tethys.* An Aquatic Biogeography of the Levant. 1989
    ISBN 0-7923-0189-7
64. B. R. Allanson, R. C. Hart, J. H. O'Keeffe and R. D. Robarts: *Inland Waters of Southern Africa.* An Ecological Perspective. 1990   ISBN 0-7923-0266-4
65. F. di Castri, A. J. Hansen and M. Debussche (eds.): *Biological Invasions in Europe and the Mediterranean Basin.* 1990   ISBN 0-7923-0411-X
66. R. W. Edwards, A. S. Gee and J. H. Stoner (eds.): *Acid Waters in Wales.* 1990
    ISBN 0-7923-0493-4
67. S. K. Jain and L. W. Botsford (eds.): *Applied Populations Biology.* 1992
    ISBN 0-7923-1425-5
68. C. Dejoux and A. Iltis (eds.): *Lake Titicaca* [Peru/Bolivia]. A Synthesis of Limnological Knowledge. 1992   ISBN 0-7923-1663-0
69. R.B. Wood and R.V. Smith (eds.): *Lough Neagh.* The Ecology of a Multipurpose Water Resource. 1993   ISBN 0-7923-2112-X
70. P.E. Ouboter (ed.): *The Freshwater Ecosystems of Suriname.* 1993
    ISBN 0-7923-2408-0
71. M.A. Brunt and J.E. Davies (eds.): *The Cayman Islands.* Natural History and Biogeography. 1994   ISBN 0-7923-2462-5
72. V. Fet and K. Atamuradov (eds.): *Biogeography and Ecology of Turkmenistan.* 1994
    ISBN 0-7923-2738-1
73. F.D. Por: *The Pantanal of Mato Grosso (Brazil).* World's Largest Wetlands. 1995
    ISBN 0-7923-3481-7